BARRON'S
EARLY ACHIEVER

GRADE 2

MATH WORKBOOK
ACTIVITIES & PRACTICE

REVIEW · UNDERSTAND · DISCOVER

Published by Kaplan North America, LLC, d/b/a Barron's Educational Series
1515 West Cypress Creek Road
Fort Lauderdale, Florida 33309
www.barronseduc.com

ISBN 978-1-5062-8137-7

10 9 8 7 6 5 4 3 2 1

Kaplan North America, LLC, dba Barron's Educational Series print books are available at special quantity
discounts to use for sales promotions, employee premiums, or educational purposes. For more information or to purchase books, please call the Simon & Schuster special sales department at 866-506-1949.

Photo Credits: Page 2 © Andresr/Shutterstock, Page 4 © Clip Art/Shutterstock, Page 6 © Anita Potter/Shutterstock, Page 8 © Christine Krahl/
Shutterstock, Page 31 © (girl) Axel Alvarez/Shutterstock, Page 33 © hin255/Shutterstock, Page 46 © Vuk Varuna/Shutterstock, Page 49 © (shells)
All-stock-photos/Shutterstock, Page 67 © (clock) Ivan Lukyanchuk/Shutterstock; (digital clock) 4zevar/Shutterstock, Page 69 © (girl eating
breakfast) Nataliya Dolotko/Shutterstock; (boy sleeping) Matthew Cole/Shutterstock, Page 71 © (coins) Vladimir Wrangel/Shutterstock;
(dollar front) 236808625/Shutterstock; (dollar back) YamabikaY/Shutterstock, Page 76 © (crayon) Lucie Lang/Shutterstock; (yardstick) Bertold
Werkmann/Shutterstock; (tape) AlexRoz/Shutterstock; (door) Kir_Prime/Shutterstock, Page 77 © (pen) Julia Ivantsova/Shutterstock; (nail)
RomboStudio/Shutterstock; (paintbrush) Marek Walica/Shutterstock; (pencil) Carolyn Franks/Shutterstock; (key) pukach/Shutterstock;
(marker) luckyraccoon/Shutterstock; (books) Luis Henry Marcelo Castro/Shutterstock, Page 78 © (cell phone) Nik Merkulov/Shutterstock;
(water bottle) Betacam-SP/Shutterstock, Page 79 © (screwdriver) Volovik Andrey/Shutterstock; (thumb) NatUlrich/Shutterstock, Page 80 ©
(leaf) BluezAce/Shutterstock; (fork) Pitju/Shutterstock; (necklace) Solomin Andrey/Shutterstock, Page 81 © (toothbrush) Steven Schremp/
Shutterstock, Page 83 © (blue pencil) LUMPANG, MOONMUANG/Shutterstock; (red pencil) Garsya/Shutterstock; (licorice) Markus Alison/
Shutterstock, Page 84 © AlexeyZet/Shutterstock, Page 85 © bonchan/Shutterstock, Page 87 © Andy Dean Photography/Shutterstock, Page 89
© (left cookies) Sathit/Shutterstock; (right cookies) M. Unal Ozmen/Shutterstock, Page 90 © 122940100/Shutterstock, Page 96 © (comb)
Nemika_Polted/Shutterstock; (chocolate bar) Gingko/Shutterstock; (tube) Maxx-Studio/Shutterstock, Page 97 © (corn) larryrains/Shutterstock;
© Maxi_m/Shutterstock, Page 108 © TheBlackRhino/Shutterstock, Page 129 © Zern Liew/Shutterstock, Page 130 © Aliaksei_7799/Shutterstock,
Page 132 © (left pizza) Elisanth/Shutterstock; hidesy/Shutterstock

Introduction

Barron's Early Achiever workbooks are based on sound educational practices and include both parent-friendly and teacher-friendly explanations of specific learning goals and how students can achieve them through fun and interesting activities and practice. This exciting series mirrors the way mathematics is taught in the classroom. *Early Achiever Grade 2 Math* presents these skills through different units of related materials that reinforce each learning goal in a meaningful way. The Review, Understand, and Discover sections assist parents, teachers, and tutors in helping students apply skills at a higher level. Additionally, students will become familiar and comfortable with the manner of presentation and learning, as this is what they experience every day in the classroom. These factors will help early achievers master the skills and learning goals in math and will also provide an opportunity for parents to play a larger role in their children's education.

Introduction to Problem Solving and Mathematical Practices

This book will help to equip both students and parents with strategies to solve math problems successfully. Problem solving in the mathematics classroom involves more than calculations alone. It involves a student's ability to consistently show his or her reasoning and comprehension skills to model and explain what he or she has been taught. These skills will form the basis for future success in meeting life's goals. Working through these skills each year through the twelfth grade sets the necessary foundation for collegiate and career success. Students will be better prepared to handle the challenges that await them as they gradually enter into the global marketplace.

Making Sense of the Problem-Solving Process

For students: It is important that you be able to make sense of word problems, write word problems with numbers and symbols, and be able to prove when you are right as well as to know when a mistake happened. You may solve a problem by drawing a model or using a chart, list, or other tool. When you get your correct answer, you must be able to explain how and why you chose to solve it that way. Every word problem in this workbook allows you to practice these skills, helping to prepare you for the demands of problem solving in your second-grade classroom. The first unit of this book discusses the **Ace It Time!** section of each lesson. **Ace It Time!** will help you master these practices.

While Doing Mathematics You Will...

1. Make sense of problems and become a champion in solving them

- Solve problems and discuss how you solved them
- Look for a starting point and plan to solve the problem
- Make sense (meaning) of a problem and search for solutions
- Use concrete objects or pictures to solve problems
- Check over work by asking, "Does this make sense?"
- Plan out a problem-solving approach

2. Reason on concepts and understand that they are measurable

- Understand how numbers represent specific quantities
- Connect quantities to written symbols
- Take a word problem and represent it with numbers and symbols
- Know and use different properties of operations
- Connect addition and subtraction to length

3. Construct productive arguments and compare the reasoning of others

- Construct arguments using concrete objects, pictures, drawings, and actions
- Practice having conversations/discussions about math
- Explain your own thinking to others and respond to the thinking of others
- Ask questions to clarify the thinking of others (How did you get that answer? Why is that true?)
- Justify your answer and determine if the thinking of others is correct

4. Model with mathematics

- Determine ways to represent the problem mathematically
- Represent story problems in different ways; examples may include numbers, words, drawing pictures, using objects, acting out, making a chart or list, writing equations
- Make connections between different representations and explain
- Evaluate your answers and think about whether or not they make sense

5. Use appropriate tools strategically

- Consider available tools when solving math problems
- Choose tools appropriately
- Determine when certain tools might be helpful
- Use technology to help with understanding

6. Attend to detail

- Develop math communication skills by using clear and exact language in your math conversations
- Understand meanings of symbols and label appropriately
- Calculate accurately

7. Look for and make use of structure

- Apply general math rules to specific situations
- Look for patterns or structure to help solve problems
- Adopt mental math strategies based on patterns, such as making ten, fact families, and doubles

8. Look for and express regularity in repeated reasoning

- Notice repeated calculations and look for shortcut methods to solve problems (for example, rounding up and adjusting the answer to compensate for the rounding)
- Evaluate your own work by asking, "Does this make sense?"

Contents

Contents

Mathematical Foundations for Grade 2

Problem-Solving Concepts

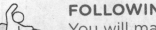

FOLLOWING THE OBJECTIVE
You will make sense of word problems and use strategies to solve them.

LEARN IT: You might be asked to work on difficult problems with at least two steps. When solving these problems, you should first read the problem and understand what the problem is asking you to do. You should also be able to explain your thinking and use the correct vocabulary. Now let's get started on the problem-solving process!

STEP 1: UNDERSTAND

What's the Question?

Math problems in second grade can have several steps. Each step is a task on our rubric. First, read the problem and ask yourself, "What question do I have to answer?" and "Will it take more than one step to solve the problem?"

During each **Ace It Time!** section, your first step on the rubric is to find the question that you have to answer and underline it.

		yes	no
★ ACE IT TIME!	**Did you underline the question in the word problem?**	yes ○	no ○
	Did you circle the numbers or number words?	yes ○	no ○
	Did you box the clue words that tell you what operation to use?	yes ○	no ○
	Did you use a picture to show your thinking?	yes ○	no ○
	Did you label your numbers and your picture?	yes ○	no ○
	Did you explain your thinking and use math vocabulary words in your explanation?	yes ○	no ○

PRACTICE: Underline the question.

Example: Samantha finds 32 sand dollars at the beach. Jayna finds 17 sand dollars. How many total sand dollars were found at the beach?

Ask "What question do I have to answer?" *How many total sand dollars were found at the beach?* Once you identify the question, underline it. Will it take more than one step to solve the problem? **No.**

STEP 2: IDENTIFY

What Numbers or Words Are Needed?

It is very important to locate the numbers in your story problem that will help you to solve the problem. Some problems use the word form and the standard form of a number. After you find the numbers in your story problem, circle them. Let's practice now.

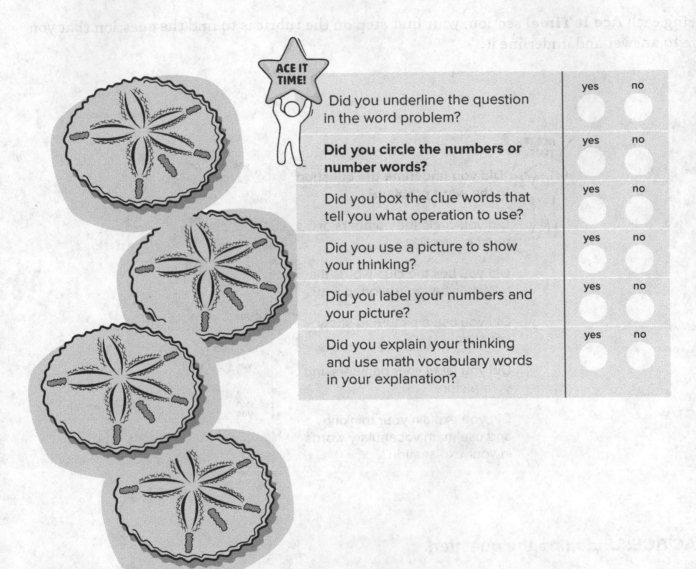

ACE IT TIME!

	yes	no
Did you underline the question in the word problem?	◯	◯
Did you circle the numbers or number words?	◯	◯
Did you box the clue words that tell you what operation to use?	◯	◯
Did you use a picture to show your thinking?	◯	◯
Did you label your numbers and your picture?	◯	◯
Did you explain your thinking and use math vocabulary words in your explanation?	◯	◯

PRACTICE: Circle the numbers.

Example: Samantha finds ⟲32⟲ sand dollars at the beach. Jayna finds ⟲17⟲ sand dollars. <u>How many total sand dollars were found at the beach?</u>

STEP 3: RECOGNIZE THE SUPPORTING DETAILS

Name the Operation.

Now let's find the supporting details or important words that will help you solve the problem. There are Samantha's 32 sand dollars and Jayna's 17 sand dollars. You need to find the total number of sand dollars. Put a box around those helpful words.

ACE IT TIME!

	yes	no
Did you underline the question in the word problem?	○	○
Did you circle the numbers or number words?	○	○
Did you box the clue words that tell you what operation to use?	○	○
Did you use a picture to show your thinking?	○	○
Did you label your numbers and your picture?	○	○
Did you explain your thinking and use math vocabulary words in your explanation?	○	○

PRACTICE: Put a box around the clues.

Example: Samantha finds 32 sand dollars at the beach. Jayna finds 17 sand dollars. How many total sand dollars were found at the beach?

Think: Are you using addition, subtraction, or both of these operations to solve this problem? How do you know? You will use only addition to solve this problem. The supporting details in the problem help you figure this out. The problem wants the total number of *both* girls' sand dollars.

STEPS 4–5: SOLVE AND LABEL

It is important that you connect words in your problem to pictures and numbers. Before solving, you should draw a picture or write a math equation to solve the problem. Make sure to label your pictures and equations. Then check "Yes" on the checklist.

PRACTICE: Draw sand dollars.

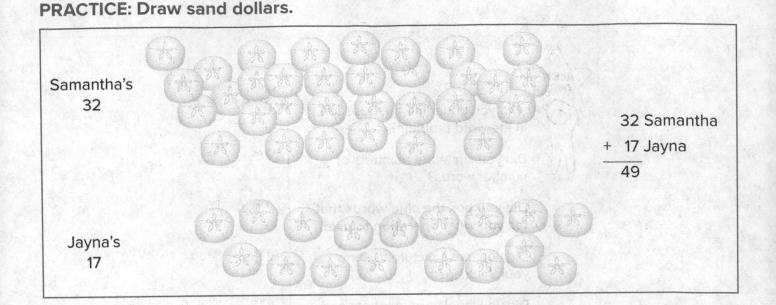

Samantha's
32

Jayna's
17

32 Samantha
+ 17 Jayna
―――
49

![ACE IT TIME!]

	yes	no
Did you underline the question in the word problem?	○	○
Did you circle the numbers or number words?	○	○
Did you box the clue words that tell you what operation to use?	○	○
Did you use a picture to show your thinking?	○	○
Did you label your numbers and your picture?	○	○
Did you explain your thinking and use math vocabulary words in your explanation?	○	○

STEP 6: EXPLAIN

Write a Response. Use Math Vocabulary.

You are almost done! Now explain your answer and show your thinking. Write in complete sentences to explain the steps you used to solve the problem. Make sure you use the vocabulary words in the Math Vocabulary box to help you!

PRACTICE: Explain your answer.

Example: Samantha finds 32 sand dollars at the beach. Jayna finds 17 sand dollars. How many total sand dollars were found at the beach?

Explanation: I knew I needed to find the **total** number of sand dollars. First, I drew 32 sand dollars for Samantha. Then I drew 17 **more** sand dollars for Jayna. I said 32 in my head and **counted** 17 **more** to get the **sum** of 49. So, 32 + 17 = 49.

ACE IT TIME!

Math Vocabulary

total

sum

more

counted

equal

	yes	no
Did you underline the question in the word problem?		
Did you circle the numbers or number words?		
Did you box the clue words that tell you what operation to use?		
Did you use a picture to show your thinking?		
Did you label your numbers and your picture?		
Did you explain your thinking and use math vocabulary words in your explanation?		

UNIT 2

Number Concepts

Odd or Even?

FOLLOWING THE OBJECTIVE
You will know if a number is even or odd.

LEARN IT: Every number is either *even* or *odd*. You can use a *ten-frame* to help you. The ten-frame will help you divide the total into groups of two to see if each number makes a *pair* or if there is one left over.

Example 1: Jack has 5 buttons. Does he have an even or odd amount of buttons?

Show the number 5 in the ten-frame:
The two **red** buttons make a pair.
The two **blue** buttons make a pair.
The one **green** button is left over.

Jack has an odd number of buttons because 5 is an odd number.

Example 2: Kelli has 8 buttons. Does she have an even or odd amount of buttons?

Show the number 8 in the ten-frame:
There are 4 pairs: a **red**, a **blue**, a **green**, and a **brown** pair.
There are no buttons left over.

Kelli has an even number of buttons because 8 is an even number.

Example 3: Mike has 17 buttons. Does he have an even or odd amount of buttons?

 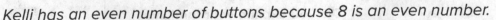

17 is an odd number because there is a button without a pair.

PRACTICE: Now you try Show the numbers in the ten-frame boxes by making pairs. Then circle if the number is even or odd.

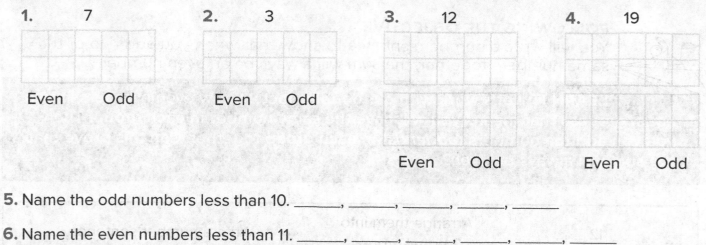

1. 7

    Even   Odd

2. 3

    Even   Odd

3. 12

    Even   Odd

4. 19

    Even   Odd

5. Name the odd numbers less than 10. _____, _____, _____, _____, _____

6. Name the even numbers less than 11. _____, _____, _____, _____, _____, _____

Marci was having a birthday party. Her mother said she could have an *even* number of friends over for the party, but no more than 13 friends. What is the greatest number of friends she could have over? Show your work and explain your thinking on a piece of paper.

Math Vocabulary

even

odd

pairs

ten-frame

pair

ACE IT TIME!

	yes	no
Did you underline the question in the word problem?		
Did you circle the numbers or number words?		
Did you box the clue words that tell you what operation to use?		
Did you use a picture to show your thinking?		
Did you label your numbers and your picture?		
Did you explain your thinking and use math vocabulary words in your explanation?		

Math on the Move   Make two columns on a piece of paper. Name one column "Even" and one column "Odd." Write the numbers from 1 to 20 in the correct columns.

Equal Addends and Even Numbers

FOLLOWING THE OBJECTIVE
You will write a number sentence to show that when you add two of the same numbers together, the *sum* will always be an even number.

LEARN IT: Let's work with our doubles facts! Show even numbers in two *equal* groups. That means you will have the same number in each group. Then you can add the numbers *(addends)* in each group together. Can this prove that the sum will always be *even*?

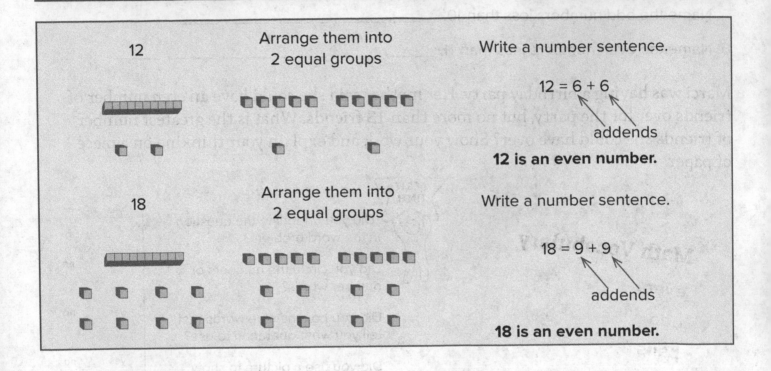

12 — Arrange them into 2 equal groups — Write a number sentence.

$12 = 6 + 6$

addends

12 is an even number.

18 — Arrange them into 2 equal groups — Write a number sentence.

$18 = 9 + 9$

addends

18 is an even number.

PRACTICE: Now you try

Add equal numbers to get an even number for the sum. Shade in the ten-frames to show your work. Use different colors.

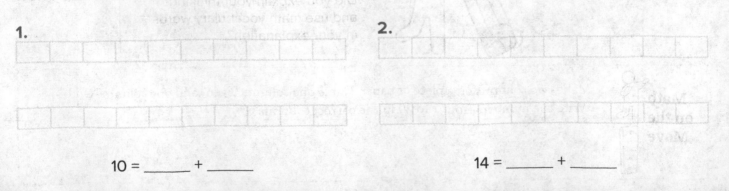

1.

$10 = \underline{\hspace{1cm}} + \underline{\hspace{1cm}}$

2.

$14 = \underline{\hspace{1cm}} + \underline{\hspace{1cm}}$

3.

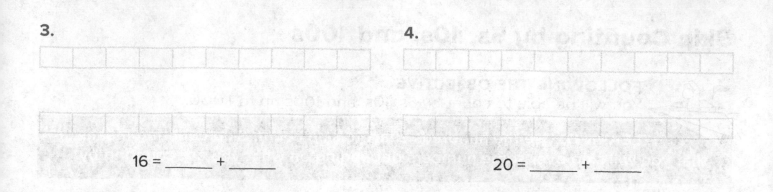

16 = _____ + _____

4.

20 = _____ + _____

The PE teacher at Sunnyville Elementary has 30 Hula-Hoops. She has two colors of hoops and the same number of each color. Her first class has 14 girls and 16 boys. Can every girl get the same color? Can every boy get the other color?

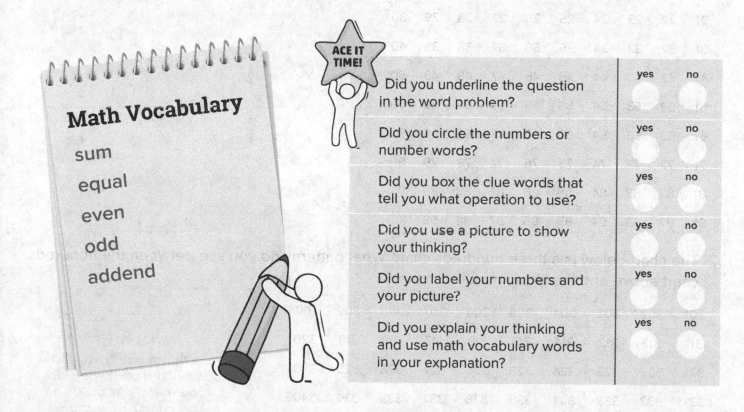

Math Vocabulary

sum

equal

even

odd

addend

ACE IT TIME!

	yes	no
Did you underline the question in the word problem?	○	○
Did you circle the numbers or number words?	○	○
Did you box the clue words that tell you what operation to use?	○	○
Did you use a picture to show your thinking?	○	○
Did you label your numbers and your picture?	○	○
Did you explain your thinking and use math vocabulary words in your explanation?	○	○

Math on the Move

Write as many number sentences as you can that add the same two numbers. Try to find any sums that are not even.

Skip Counting by 5s, 10s, and 100s

FOLLOWING THE OBJECTIVE
You will be able to count by 5s, 10s, and 100s up to 1,000.

LEARN IT: Use the hundreds chart to *skip count*. Skip counting means that you will skip over some numbers depending on what you are counting by.

1. Use the hundreds chart below to skip count by 5s. Circle all the numbers you land on. Now use the chart to skip count by 10s. Draw an "X" on all the numbers you land on.

1	2	3	4	5	6	7	8	9	10
11	12	13	14	15	16	17	18	19	20
21	22	23	24	25	26	27	28	29	30
31	32	33	34	35	36	37	38	39	40
41	42	43	44	45	46	47	48	49	50
51	52	53	54	55	56	57	58	59	60
61	62	63	64	65	66	67	68	69	70
71	72	73	74	75	76	77	78	79	80
81	82	83	84	85	86	87	88	89	90
91	92	93	94	95	96	97	98	99	100

2. The chart below is a three hundreds chart. What patterns do you see between the hundreds chart above and this three hundreds chart?

301	302	303	304	305	306	307	308	309	310
311	312	313	314	315	316	317	318	319	320
321	322	323	324	325	326	327	328	329	330
331	332	333	334	335	336	337	338	339	340
341	342	343	344	345	346	347	348	349	350
351	352	353	354	355	356	357	358	359	360
361	362	363	364	365	366	367	368	369	370
371	372	373	374	375	376	377	378	379	380
381	382	383	384	385	386	387	388	389	390
391	392	393	394	395	396	397	398	399	400

think!
What numbers would you land on if you skip counted by 5s or 10s on this chart? How can we use a chart like this to skip count by 100s?

PRACTICE: Now you try

Fill in the missing numbers.

1. 10, _____, 30, 40, _____, 60, _____, _____, 90, 100

2. _____, _____, 15, 20, _____, _____, 35, _____, 45, _____, _____

3. 550, _____, _____, 565, _____, 575, _____

4. 870, _____, _____, _____, _____, 920, _____

5. 100, 200, _____, 400, _____, _____, 700

6. Tell the pattern you see when you count by 5s.

> **think!**
> How can you use a number chart to help you figure out what you are counting by?

Alexis wanted to count backward from 356 by fives until she got to 291. What numbers would she say and in what order? Show your work and explain your thinking on a piece of paper.

ACE IT TIME!

	yes	no
Did you underline the question in the word problem?	◯	◯
Did you circle the numbers or number words?	◯	◯
Did you box the clue words that tell you what operation to use?	◯	◯
Did you use a picture to show your thinking?	◯	◯
Did you label your numbers and your picture?	◯	◯
Did you explain your thinking and use math vocabulary words in your explanation?	◯	◯

Math Vocabulary

skip counting

hundreds chart

Math on the Move

Use a set of number cards. Flip a card over and use that number as the starting number. Count by 2s for ten numbers. Count by 5s for ten numbers. Count by 10s for ten numbers.
Example: If you turn over a 3:
 Count by twos: 3, 5, 7, 9, 11, 13, 15, 17, 19, 21
 Count by fives: 3, 8, 13, 18, 23, 28, 33, 38, 43, 48
 Count by tens: 3, 13, 23, 33, 43, 53, 63, 73, 83, 93

Show Three-Digit Numbers with Base-Ten Blocks

FOLLOWING THE OBJECTIVE
You will explain and demonstrate the meaning of each digit in a three-digit number by using pictures of groups of hundreds, tens, and ones.

LEARN IT: The order of the digits within a number tells you the *value* of the number. Each number can be shown with base-ten blocks.

Example: Show why the number 26 is different than 62.

In a two-digit number, the first digit tells you how many groups of ten are in the number. The second digit tells you how many ones. So 26 has a value that is less than 62. It has only 2 tens while 62 has 6 tens.

Now think about a three-digit number. 465 has three digits. The 4 tells you how many groups of hundreds are in the number.

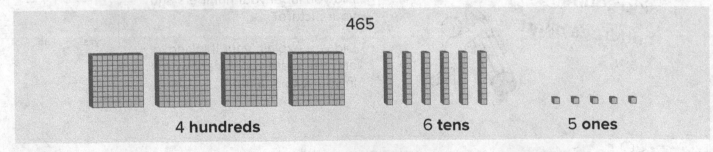

In a three-digit number, the first digit tells you how many groups of hundred. The second digit tells you how many groups of ten, and the third digit tells you how many ones. 465 has a value much bigger than 26 and 62 because it has groups of hundreds.

PRACTICE: Now you try

Draw base-ten blocks ☐ ▯ ☐ to show the value of the digits in each number.

1. 45 **2.** 369 **3.** 701 **4.** 214 **5.** 503

Circle the digit in the tens place.

6. 49 **7.** 84 **8.** 154 **9.** 805 **10.** 667

Circle the digit in the hundreds place.

11. 251 **12.** 739 **13.** 893 **14.** 340 **15.** 2,450

Jordan thinks 204 is greater than 240. Can you explain to her using place value why she is incorrect? Show your work and explain your thinking on a piece of paper.

ACE IT TIME!

Math Vocabulary

value

digit

ones

tens

hundreds

	yes	no
Did you underline the question in the word problem?	yes	no
Did you circle the numbers or number words?	yes	no
Did you box the clue words that tell you what operation to use?	yes	no
Did you use a picture to show your thinking?	yes	no
Did you label your numbers and your picture?	yes	no
Did you explain your thinking and use math vocabulary words in your explanation?	yes	no

Math on the Move

Think of money and place value. What money could you use to show the hundreds block? What coin could you use for the tens block? What coin could you use for the ones block?

Expanded and Word Forms

FOLLOWING THE OBJECTIVE
You will use what you know about the value of the digits in a number to write numbers in expanded form and by using number words.

LEARN IT: Let's look at the numbers 18, 61, and 236. Think about what the digits in each number mean. We can show the number four different ways like in the chart below.

Number	Number Words	Base-Ten Picture	Expanded Form
18	eighteen	1 ten 8 ones 10 8	10 + 8 = 18
61	sixty-one	6 tens 1 60 1	60 + 1 = 61
236	two hundred thirty-six	2 hundreds 3 tens 6 ones 200 30 6	200 + 30 + 6 = 236

Use this chart to help spell out number words:

Ones		Teens		Tens	
zero	0	eleven	11	ten	10
one	1	twelve	12	twenty	20
two	2	thirteen	13	thirty	30
three	3	fourteen	14	forty	40
four	4	fifteen	15	fifty	50
five	5	sixteen	16	sixty	60
six	6	seventeen	17	seventy	70
seven	7	eighteen	18	eighty	80
eight	8	nineteen	19	ninety	90
nine	9				

When you write a two-digit number, you use a word from the "Tens" column and a word from the "Ones" column. Be sure to use a hyphen between the tens and ones!

34 = thirty-four   95 = ninety-five   22 = twenty-two   843 = eight hundred forty-three

PRACTICE: Now you try

1. Draw a line from the number in the first column to the matching number form in the second and third columns.

21	fifty-seven	600 + 70 + 8
57	three hundred forty-two	400 + 8
408	twenty-one	50 + 7
342	six hundred seventy-eight	300 + 40 + 2
678	four hundred eight	20 + 1

Write the expanded form for the following numbers and then write the words.

2. 729 _____ + _____ + _____ _____

3. 503 _____ + _____ + _____ _____

4. 956 _____ + _____ + _____ _____

Why do you need to use the zero in the number *one hundred four*? Use what you know about the different ways to write a number. Show your work and explain your thinking on a piece of paper.

ACE IT TIME!

	yes	no
Did you underline the question in the word problem?	○	○
Did you circle the numbers or number words?	○	○
Did you box the clue words that tell you what operation to use?	○	○
Did you use a picture to show your thinking?	○	○
Did you label your numbers and your picture?	○	○
Did you explain your thinking and use math vocabulary words in your explanation?	○	○

Math Vocabulary

expanded form

hyphen

Math on the Move

Write your birthday date using number words. Write your address using number words. Look in the newspaper and circle any number words you find. Write them in both number and expanded forms.

Show Numbers Different Ways

FOLLOWING THE OBJECTIVE
You will use what you know about the value of a number to show numbers in different ways.

LEARN IT: You can show or make numbers in different ways. Numbers can be shown as ones or you can make groups of ten.

Example: Let's look at the number 23.

One Way	Another Way	A Third Way
Tens Ones	Tens Ones	Tens Ones
2 tens + 3 ones	1 ten + 13 ones	0 tens + 23 ones
20 + 3	10 + 13	0 + 23
23	23	23

PRACTICE: Now you try

Show the number 48 in three different ways:

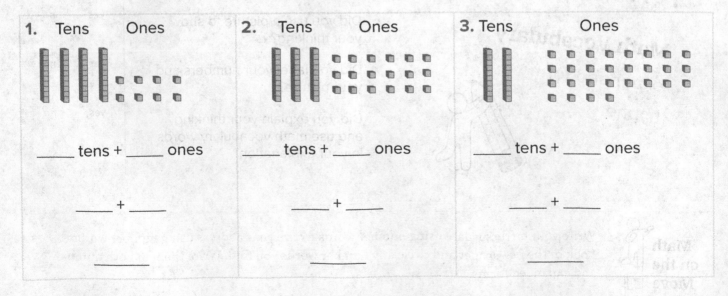

1. Tens   Ones

_____ tens + _____ ones

_____ + _____

_____

2. Tens   Ones

_____ tens + _____ ones

_____ + _____

_____

3. Tens   Ones

_____ tens + _____ ones

_____ + _____

_____

Show the number 54 in three different ways:

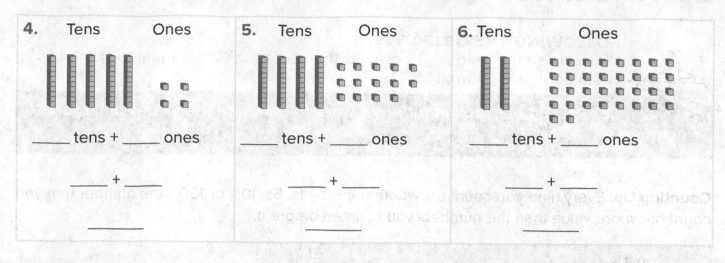

4. Tens   Ones

_____ tens + _____ ones

_____ + _____

_____

5. Tens   Ones

_____ tens + _____ ones

_____ + _____

_____

6. Tens   Ones

_____ tens + _____ ones

_____ + _____

_____

Rob has to buy 33 cupcakes for his class party. He can buy them as singles or in boxes of 10. Show all of the different ways that Rob can buy cupcakes. Show your work and explain your thinking on a piece of paper.

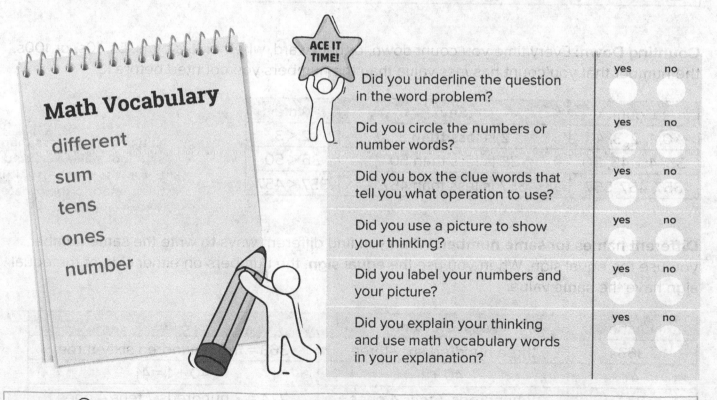

Math Vocabulary

different

sum

tens

ones

number

ACE IT TIME!

	yes	no
Did you underline the question in the word problem?		
Did you circle the numbers or number words?		
Did you box the clue words that tell you what operation to use?		
Did you use a picture to show your thinking?		
Did you label your numbers and your picture?		
Did you explain your thinking and use math vocabulary words in your explanation?		

Math on the Move

Change lying around your home can be a good learning tool and a great way to practice showing different ways to make a number. Practice making change in different ways. For example, you could make 25 cents using a quarter, two dimes and a nickel, or 25 pennies.

Compare Three-Digit Numbers

FOLLOWING THE OBJECTIVE
You will use the math signs greater than, less than, and equal to (>, <, =) when comparing numbers.

LEARN IT: Let's look at different ways to compare the values of larger numbers.

Counting Up: Every time you count up, whether it is by 1s, 5s, 10s, or 100s, the number that you count has *more* value than the numbers you counted before it.

Count	Say	Write
0, 1, 2, 3, 4	3 is **greater than** 2	3 > 2
35, 40, 45, 50	50 is greater than 35	50 > 35
357, 457, 557	457 is greater than 357	457 > 357

think!
The greater than sign "opens" to the greater number!

Counting Down: Every time you count down, or backward, whether it is by 1s, 5s, 10s, or 100s, the number that you count has *less* value than the numbers you counted before it.

Count	Say	Write
0, 1, 2, 3, 4	2 is **less than** 3	2 < 3
35, 40, 45, 50	35 is less than 50	35 < 50
357, 457, 557	357 is less than 457	357 < 457

think!
Notice that the less than sign looks like a slanted "L" for less than! Also, it is still opened to the greater number.

Different names for same number: When you find different ways to write the same number, you use the equal sign. When you use the **equal sign**, the numbers on either side of the equal sign have the same value.

Number	Say New Name	Write
363	three hundred sixty-three	363 = three hundred sixty-three
41	40 + 1	40 + 1 = 41
274	2 hundreds 7 tens 4 ones	274 = 2 hundreds 7 tens 4 ones
56	4 tens 16 ones	56 = 4 tens 16 ones

PRACTICE: Now you try

Put the correct sign, <, >, or =, in the circles below. Write how you would say it.

1. 673 ◯ 798 _____

2. 453 ◯ 443 _____

3. 305 ◯ 30 + 5 _____

4. 599 ◯ 299 _____

> **think!**
> What do you know about place value? Use the location of the digits to help you decide.

Use what you know about comparing numbers to answer the questions below.

5. If you have two three-digit numbers, would you first look at the hundreds place or the ones place first to compare the numbers? _____

6. When comparing two three-digit numbers and the hundreds digits in both numbers are the same, what place value would you look at next to compare which number is smaller?

_____

7. Put the following numbers in order from least to greatest.

463, 75, 189, 989, 198, 364, 104

_____

Joey thought he had more pennies than Joan. He had 101 pennies and Joan had 110. Where did Joey make a mistake? Show your work and explain your thinking on a piece of paper.

ACE IT TIME!

	yes	no
Did you underline the question in the word problem?	◯	◯
Did you circle the numbers or number words?	◯	◯
Did you box the clue words that tell you what operation to use?	◯	◯
Did you use a picture to show your thinking?	◯	◯
Did you label your numbers and your picture?	◯	◯
Did you explain your thinking and use math vocabulary words in your explanation?	◯	◯

Math Vocabulary

place value

greater than

less than

equal to

Math on the Move

Roll three number cubes. Make a three-digit number. Roll the cubes again. Make another three-digit number. Write a number sentence using <, >, or = to make the number sentence true.

Mentally Add or Subtract 10 or 100

FOLLOWING THE OBJECTIVE
You will add and subtract 10 or 100 to a number without having to use a picture or a number sentence.

LEARN IT: It is easy to add or subtract tens to a number in your head if you look for patterns. Look to see how the digit in the tens place changes when tens are added or subtracted in numbers.

Hundreds	Tens	Ones	Add or Subtract	=	Hundreds	Tens	Ones
	3	9	+ 1 ten (10)	=		**4**	9
	3	2	+ 5 tens (50)	=		**8**	2
6	**7**	4	− 4 tens (40)	=	6	**3**	4
4	**0**	5	+ 1 ten (10)	=	4	**1**	5
8	**8**	8	− 6 tens (60)	=	8	**2**	8

Now look to see how the digit in the hundreds place changes when hundreds are added or subtracted in numbers.

Hundreds	Tens	Ones	Add or Subtract	=	Hundreds	Tens	Ones
	3	9	+ 1 hundred (100)	=	**1**	3	9
	3	2	+ 5 hundreds (500)	=	**5**	3	2
6	7	4	− 4 hundreds (400)	=	**2**	7	4
4	0	5	+ 1 hundred (100)	=	**5**	0	5
8	8	8	− 6 hundreds (600)	=	**2**	8	8

If you know the basic facts to 10, adding or subtracting 10s and 100s in your head will be easy for you to do.

PRACTICE: Now you try
Add or subtract tens to the following numbers.

1. 23
+ 1 ten
_____

2. 45
− 3 tens
_____

3. 88
− 4 tens
_____

4. 566
+ 2 tens
_____

5. 703
+ 5 tens
_____

6. 1,045
− 3 tens
_____

Add or subtract hundreds to the following numbers.

7. 72
+ 1 hundred
_____

8. 67
+ 3 hundreds
_____

9. 703
– 7 hundreds
_____

10. 493
+ 2 hundreds
_____

11. 992
– 5 hundreds
_____

12. 1,367
– 3 hundreds
_____

Fill in the missing numbers. *Hint:* Look for a pattern!

13. 41, 51, 61, _____, 81, _____, _____, _____, 121

14. 1, 101, 201, _____, 401, _____, _____, _____, 801

15. 547, 557, _____, _____, 587, _____, 607, _____

Lucy walked 665 steps during PE. Mark walked 365 steps. How many more hundreds of steps did Lucy walk than Mark? Explain how you know. Show your work and explain your thinking on a piece of paper.

ACE IT TIME!

Math Vocabulary

tens

hundreds

digit

place value

	yes	no
Did you underline the question in the word problem?	○	○
Did you circle the numbers or number words?	○	○
Did you box the clue words that tell you what operation to use?	○	○
Did you use a picture to show your thinking?	○	○
Did you label your numbers and your picture?	○	○
Did you explain your thinking and use math vocabulary words in your explanation?	○	○

Math on the Move

Use a set of number cards. Take three cards and make a three-digit number. Take another card. Use that card to tell you how many 10s and 100s you can add or subtract from your first number.

REVIEW

Congratulations! You've finished the lessons for Unit 2. This means you've learned that skip counting is faster than adding over and over again. You've learned about numbers and how to write them in different forms. You've practiced comparing numbers. You can even add 10 and 100 to numbers in your mind and tell if the number is odd or even!

Now it's time to prove your number-concept skills. Solve the problems below. Use all of the methods you have learned.

Activity Section 1

Write the number shown with the base-ten blocks.

1. 2 flats + 8 rods + 2 units   The number is _____.

2. 3 flats + 4 rods + 6 units   The number is _____.

3. 4 flats + 12 rods + 5 units   The number is _____.

4. 1 flat + 15 rods + 8 units   The number is _____.

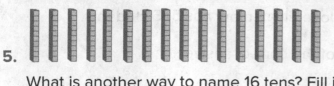

5.

What is another way to name 16 tens? Fill in the circle of the correct answer.

◯ 16 ones

◯ 1 ten and 6 ones

◯ 1 hundred 6 ones

◯ 1 hundred 6 tens

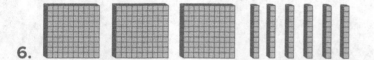

6.

Write how many tens. _____ tens

Write how many hundreds. _____ hundreds

Write the number. _____

7. Read the number and draw a quick picture. Write the number in different ways.

one hundred sixty-two

_____ hundred(s) _____ ten(s) _____ one(s)

_____ + _____ + _____

Activity Section 2

1. Sand Lake Elementary has 474 students. Forest Lakes Elementary has 447 students.

Which of the following is true? Fill in the circle of the correct answer.

◯ 474 = 447

◯ 474 < 447

◯ 474 > 447

2. San Marks Elementary has 523 students. San Jose Elementary has 535 students.

   Which of the following is true? Fill in the circle of the correct answer.

   ◯ 523 = 535

   ◯ 523 < 535

   ◯ 523 > 535

3. Look at the numbers given and find the next two numbers.

   145, 245, 345, 445, 545, _____, _____

   12, 22, 32, 42, 52, _____, _____

   25, 30, 35, 40, 45, 50, _____, _____

4. Number Riddle:

   There is a three-digit number with the digit 7 in the ones place, the digit 2 in the tens place, and the digit 4 in the hundreds place.

   What is the number? _____

   Is this number odd or even? _____

Activity Section 3

Will the sum of the following numbers be even or odd?

1. 244	2. 353	3. 568	4. 182
+ 468	+ 271	+ 137	+ 639
_____	_____	_____	_____

Even/Odd   Even/Odd   Even/Odd   Even/Odd

UNDERSTAND

Understand the meaning of what you have learned and apply your knowledge.

It is important for you to understand the value of a number and that each number can be created using many different combinations. Getting to know numbers will give you a powerful tool that you can use in many areas of math and the real world!

Activity Section

Ms. Mukerjee asked her students to make the number 53 as many ways as they could. She found that her students showed the number several different ways. After grading her students' papers, she realized that one of these ways was wrong. Which one of her students was wrong? How do you know?

Ben's Way:	Dean's Way:
5 tens and 3 ones	1 ten and 43 ones
Sarah's Way:	Donna's Way:
3 tens and 23 ones	2 tens and 43 ones
Laura's Way:	
4 tens and 13 ones	

DISCOVER

You will understand how place value can help you see and continue counting patterns. Using a number grid can help you mentally add and subtract 10 or 100 to a given number.

Activity Section

Shade the numbers in the grid below to help you answer the following questions.

601	602	603	604	605	606	607	608	609	610
611	612	613	614	615	616	617	618	619	620
621	622	623	624	625	626	627	628	629	630
631	632	633	634	635	636	637	638	639	640
641	642	643	644	645	646	647	648	649	650
651	652	653	654	655	656	657	658	659	660
661	662	663	664	665	666	667	668	669	670
671	672	673	674	675	676	677	678	679	680
681	682	683	684	685	686	687	688	689	690
691	692	693	694	695	696	697	698	699	700

1. Maria had 625 colored rubber bands that she wanted to weave into bracelets. She bought more rubber bands and now has 675 rubber bands. How many groups of 10 rubber bands did she buy?

   She bought _____ groups of 10 rubber bands.

   Write the math sentence: _____

2. Sarah had 690 colored rubber bands that she wanted to weave into bracelets. She used some rubber bands to weave bracelets and now has 630 rubber bands. How many groups of 10 rubber bands did she use?

   She used _____ groups of 10 rubber bands.

   Write the math sentence: _____

Addition and Subtraction Mental Fact Strategies

FOLLOWING THE OBJECTIVE
You will add and subtract from memory all sums of two one-digit numbers.

LEARN IT: You can use these strategies to solve *addition* problems in your head.

Count On	Change the Order	Use Doubles Facts
You can count on 1, 2, or 3.	Changing the order of the addends does not change the sum.	Use a doubles fact to find the sum of a number that has 1 more or 1 less than a doubles fact.
$7 + 1 = \underline{8}$ $7 + 2 = \underline{9}$ $7 + 3 = \underline{10}$	$7 + 3 = 10$ $3 + 7 = 10$	$6 + 7 = \underline{}$ **$6 + 6$** $+ 1 = 13$ Or you could solve it this way! **$7 + 7$** $- 1 = 13$

You can use these strategies to solve **subtraction** problems in your head.

Count Backward	think!	Count Up
$9 - 3 = 6$ Start with 9. Say **"8, 7, 6."**	Think of the addition fact: $7 - 3 = 4$ Think: $3 + 4 = 7$ So, $7 - 3 = 4$	$6 - 4 = 2$ Start with 4. Say **"5, 6."** **You counted up 2 numbers.** **That is the difference!**

PRACTICE: Now you try

Write the sums. Use the strategies learned on the previous page to solve mentally.

think!
What happens when you subtract zero from a number?

1. $5 + 1 =$ ____

   $1 + 5 =$ ____

2. $8 + 1 =$ ____

   $8 + 2 =$ ____

3. ____ $= 5 + 0$

   ____ $= 8 + 0$

4. ____ $= 9 + 9$

   ____ $= 9 + 8$

5. $6 + 7 =$ ____

   $7 + 6 =$ ____

6. $4 + 4 =$ ____

   $4 + 5 =$ ____

7. $4 + 0 =$ ____

   $7 + 0 =$ ____

8. $5 + 2 =$ ____

   $5 + 3 =$ ____

9. $8 + 2 =$ ____

   $2 + 8 =$ ____

Write the differences. Use the strategies learned on the previous page to solve mentally.

10. $7 - 1 =$ ____

    $1 +$ ____ $= 7$

11. $4 - 1 =$ ____

    $4 - 2 =$ ____

12. ____ $= 3 - 0$

    ____ $= 9 - 0$

13. $18 - 9 =$ ____

    $18 = 9 +$ ____

14. $6 - 5 =$ ____

    $6 - 4 =$ ____

15. $7 - 2 =$ ____

    $7 - 3 =$ ____

16. $6 - 0 =$ ____

    $4 - 0 =$ ____

17. $13 - 6 =$ ____

    $6 +$ ____ $= 12$

18. $8 - 2 =$ ____

    $2 +$ ____ $= 8$

Samantha drew seven pictures. Keegan drew twice as many pictures as Samantha did. How many pictures did they draw in all?

Math Vocabulary

sum

add

total

doubled

ACE IT TIME!

	yes	no
Did you underline the question in the word problem?		
Did you circle the numbers or number words?		
Did you box the clue words that tell you what operation to use?	yes no	
Did you use a picture to show your thinking?		
Did you label your numbers and your picture?		
Did you explain your thinking and use math vocabulary words in your explanation?		

Math on the Move

Play a game of "Fact War" to practice adding and subtracting. You will need a set of number cards. Deal out all cards to yourself and your partner. Keep cards face down. Each player turns over two cards and adds them together and says the sum out loud. The first player to say the sum correctly gets both sets of cards. The first player to win all the cards wins the game. You can also follow the same rules but play using subtraction!

Addition and Subtraction Fact Practice

FOLLOWING THE OBJECTIVE
You will add and subtract from memory all sums of two one-digit numbers.

LEARN IT: In this lesson, you will practice some addition and subtraction facts. You can even time yourself for fun! Remember, the more you practice, the more you will remember your facts!

PRACTICE: Now you try

Solve each problem. Watch the signs!

1. 7 + 4	2. 9 + 0	3. 8 − 3	4. 11 − 7	5. 2 + 3	6. 6 + 1
7. 2 + 2	8. 7 − 3	9. 2 + 9	10. 6 + 4	11. 0 + 2	12. 10 − 5
13. 9 + 6	14. 2 + 4	15. 3 − 1	16. 11 − 2	17. 7 − 7	18. 8 + 7
19. 2 + 5	20. 5 + 7	21. 14 − 5	22. 10 + 7	23. 9 + 5	24. 8 − 1
25. 12 − 8	26. 7 + 9	27. 8 − 0	28. 18 − 9	29. 6 + 7	30. 20 − 10
31. 15 − 9	32. 3 + 3	33. 8 + 9	34. 4 + 9	35. 13 − 5	36. 8 + 8
37. 2 + 2	38. 6 − 4	39. 5 − 3	40. 2 + 7	41. 9 + 9	42. 17 − 9

Catrina read nine pages on Monday. She read seven pages on Tuesday. There are 20 pages in her book. How many more pages does Catrina need to read to finish the book? Show your work and explain your thinking on a piece of paper.

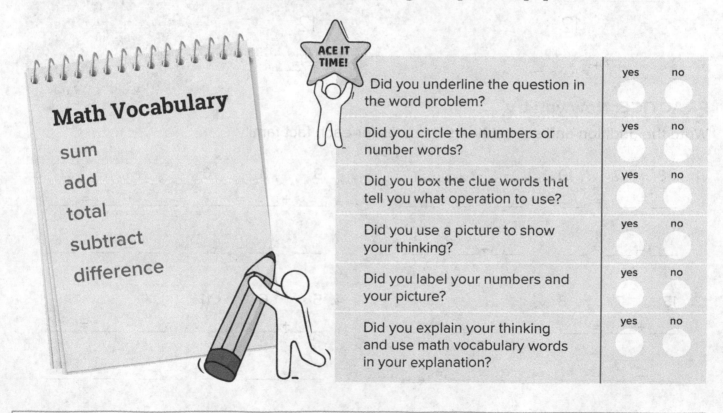

Math Vocabulary

sum

add

total

subtract

difference

ACE IT TIME!

	yes	no
Did you underline the question in the word problem?		
Did you circle the numbers or number words?		
Did you box the clue words that tell you what operation to use?		
Did you use a picture to show your thinking?		
Did you label your numbers and your picture?		
Did you explain your thinking and use math vocabulary words in your explanation?		

Math on the Move

You will need two number cubes and 24 index cards. Make two sets of number cards 1–12. Lay the cards in a row 1–12. With a partner, decide who will go first and roll both number cubes. Add the numbers together. Now turn over all the number cards that equal the sum you rolled. Your turn ends when you cannot turn over any more cards. The winner is the one who can turn over all the cards during his or her turn!

Addition and Subtraction Fact Families

FOLLOWING THE OBJECTIVE
You will add and subtract from memory all sums of two one-digit numbers.

LEARN IT: You can use addition facts to help you solve subtraction problems. Related facts have the same whole and parts. The three numbers belong together in a fact *family*.

Look at the problems below. Which three numbers belong together in this fact family? Fill in the blanks of the missing numbers.

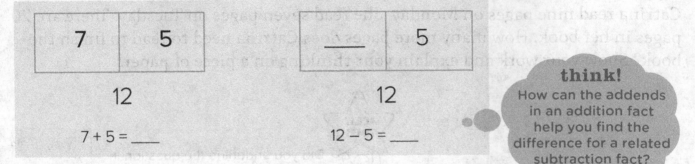

7	5

12

7 + 5 = ___

___	5

12

12 − 5 = ___

think!
How can the addends in an addition fact help you find the difference for a related subtraction fact?

PRACTICE: Now you try

Write the addition and subtraction sentences for each fact family.

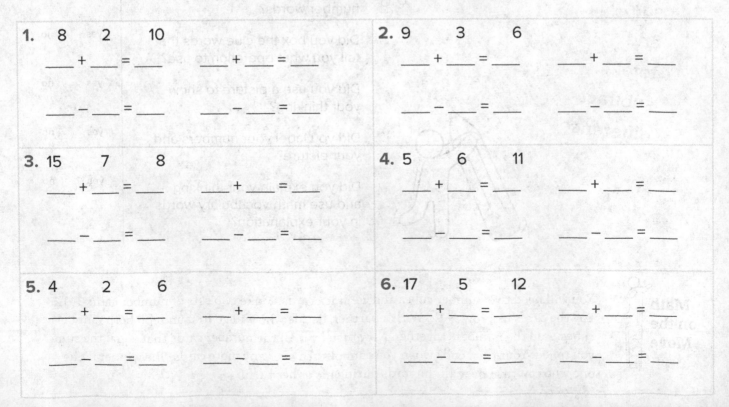

1. 8   2   10

___ + ___ = ___   ___ + ___ = ___

___ − ___ = ___   ___ − ___ = ___

2. 9   3   6

___ + ___ = ___   ___ + ___ = ___

___ − ___ = ___   ___ − ___ = ___

3. 15   7   8

___ + ___ = ___   ___ + ___ = ___

___ − ___ = ___   ___ − ___ = ___

4. 5   6   11

___ + ___ = ___   ___ + ___ = ___

___ − ___ = ___   ___ − ___ = ___

5. 4   2   6

___ + ___ = ___   ___ + ___ = ___

___ − ___ = ___   ___ − ___ = ___

6. 17   5   12

___ + ___ = ___   ___ + ___ = ___

___ − ___ = ___   ___ − ___ = ___

Write the sum and the difference for the related facts.

7. 5 + 4 = _____ 9 − 4 = _____	**8.** 7 + 4 = _____ 11 − 4 = _____	**9.** 9 + 7 = _____ 16 − 7 = _____
10. 8 + 6 = _____ 14 − 6 = _____	**11.** 3 + 5 = _____ 8 − 5 = _____	**12.** 9 + 9 = _____ 18 − 9 = _____
13. 3 + 9 = _____ 12 − 9 = _____	**14.** 4 + 6 = _____ 10 − 6 = _____	**15.** 8 + 5 = _____ 13 − 5 = _____

Mr. Milton has a bag of seven apples and a bag of eight apples. His family eats six apples. How many apples does he have left? Show your work and explain your thinking on a piece of paper.

Math Vocabulary

odd

even

sum

addend

total

ACE IT TIME!

	yes	no
Did you underline the question in the word problem?	yes	no
Did you circle the numbers or number words?	yes	no
Did you box the clue words that tell you what operation to use?	yes	no
Did you use a picture to show your thinking?	yes	no
Did you label your numbers and your picture?	yes	no
Did you explain your thinking and use math vocabulary words in your explanation?	yes	no

Math on the Move Think of an addition fact you want to remember. Write the addition fact and its related subtraction fact for that fact family. Or think of any set of three numbers and quickly say four related addition and subtraction problems for those numbers.

Addition Using Arrays

FOLLOWING THE OBJECTIVE
You will use addition to find the total number of objects in an array.

LEARN IT: You can show your understanding of story problems by drawing a picture. Let's look at the problem below.

Example: Mrs. Peter organized the desks in her classroom into three rows. There were five desks in each row. How many desks are there in Mrs. Peter's classroom?

Step 1: Draw a model showing 3 rows of 5 desks. You can use simple shapes like circles or squares as the desks.	**Step 2:** Add the rows to find the answer.
◯ ◯ ◯ ◯ ◯ ◯ ◯ ◯ ◯ ◯ ◯ ◯ ◯ ◯ ◯ *Remember:* *Rows go side to side!* There are 3 rows. There are 5 circles (desks) in each row.	5 (desks in row 1) 5 (desks in row 2) + 5 (desks in row 3) 15 total desks **think!** Could skip counting help you find the answer? Is it faster?

PRACTICE: Now you try

In the spaces below, draw a model for each problem and solve.

1. Keyera put her photographs in an album. She placed 2 pictures in a row. There were 3 rows on the page. How many pictures are on the page?	pictures _____
2. Catrina noticed there were 5 shelves on her bookcase. She put 5 books on each shelf. How many books did Catrina place on the bookcase?	books _____

3. Javeon had a sheet of postage stamps. There were 4 rows of stamps and there were 5 stamps in each row. How many stamps were there in all?

stamps _____

4. Cody's mother bought a large carton of eggs. The carton held 3 rows of eggs. There were 6 eggs in each row. How many eggs were there in all?

eggs _____

5. Inside a box of birthday candles are 2 rows of candles and 8 candles in each row. How many candles are in the box?

candles _____

Terrell had a box of crayons. There were four rows of crayons in the box. There were six crayons in each row. Two of the crayons were broken. How many crayons were not broken in Terrell's box of crayons? Show your work and explain your thinking on a piece of paper.

Math Vocabulary

total

rows

added

skip counted

ACE IT TIME!

	yes	no
Did you underline the question in the word problem?	○	○
Did you circle the numbers or number words?	○	○
Did you box the clue words that tell you what operation to use?	○	○
Did you use a picture to show your thinking?	○	○
Did you label your numbers and your picture?	○	○
Did you explain your thinking and use math vocabulary words in your explanation?	○	○

Math on the Move

Just like the problems in this lesson, you will find objects arranged in equal groups everywhere. Look around your home, your neighborhood, the mall, or even the grocery store and find as many examples as you can. Then figure out how many objects there are in that group.

Addition and Subtraction Concepts

Addition Using Place Value

FOLLOWING THE OBJECTIVE
You will add numbers using place value.

LEARN IT: Numbers can be broken apart by using their place value (100s, 10s, and 1s). You can use a number's *place value* to learn how to add larger numbers.

Example: Find the sum.

$$\begin{array}{r} 34 \\ + 53 \end{array}$$

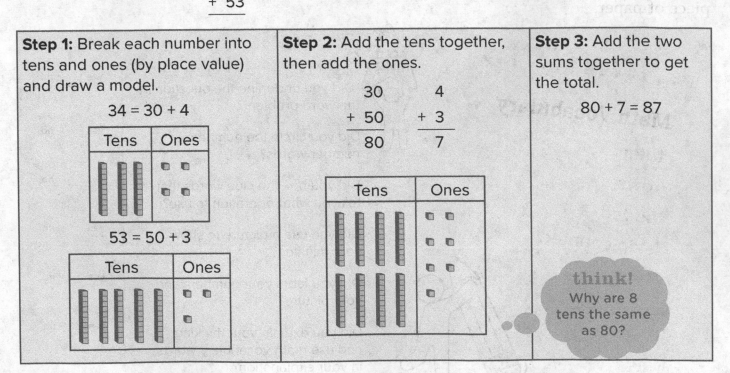

Step 1: Break each number into tens and ones (by place value) and draw a model.

34 = 30 + 4

Tens	Ones

53 = 50 + 3

Tens	Ones

Step 2: Add the tens together, then add the ones.

$$\begin{array}{r} 30 \\ + 50 \\ \hline 80 \end{array} \qquad \begin{array}{r} 4 \\ + 3 \\ \hline 7 \end{array}$$

Tens	Ones

Step 3: Add the two sums together to get the total.

80 + 7 = 87

think! Why are 8 tens the same as 80?

PRACTICE: Now you try Break each number into tens and ones. Solve.

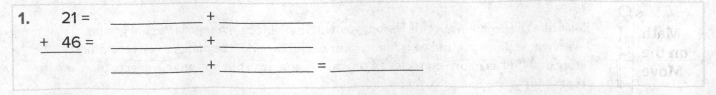

1.   21 = _____ + _____

   + 46 = _____ + _____

   _____ + _____ = _____

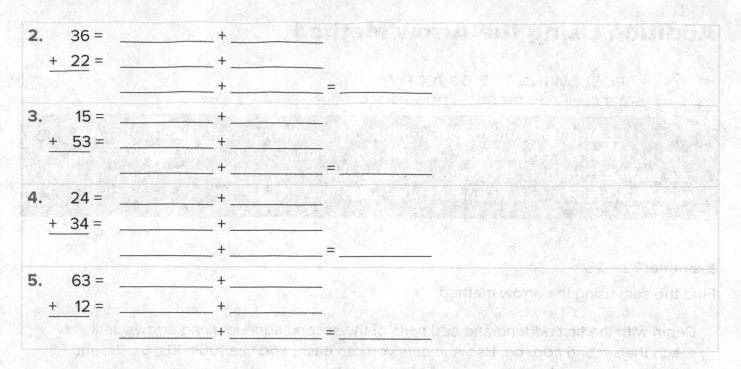

2.
$$36 = \underline{\hspace{2cm}} + \underline{\hspace{2cm}}$$
$$+ \ 22 = \underline{\hspace{2cm}} + \underline{\hspace{2cm}}$$
$$\underline{\hspace{2cm}} + \underline{\hspace{2cm}} = \underline{\hspace{2cm}}$$

3.
$$15 = \underline{\hspace{2cm}} + \underline{\hspace{2cm}}$$
$$+ \ 53 = \underline{\hspace{2cm}} + \underline{\hspace{2cm}}$$
$$\underline{\hspace{2cm}} + \underline{\hspace{2cm}} = \underline{\hspace{2cm}}$$

4.
$$24 = \underline{\hspace{2cm}} + \underline{\hspace{2cm}}$$
$$+ \ 34 = \underline{\hspace{2cm}} + \underline{\hspace{2cm}}$$
$$\underline{\hspace{2cm}} + \underline{\hspace{2cm}} = \underline{\hspace{2cm}}$$

5.
$$63 = \underline{\hspace{2cm}} + \underline{\hspace{2cm}}$$
$$+ \ 12 = \underline{\hspace{2cm}} + \underline{\hspace{2cm}}$$
$$\underline{\hspace{2cm}} + \underline{\hspace{2cm}} = \underline{\hspace{2cm}}$$

Sondra had a box of 22 crayons. Meagan gave her another box with 16 crayons. How many crayons did Sondra have in all? Show your work and explain your thinking on a piece of paper.

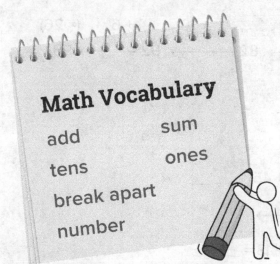

Math Vocabulary

add   sum

tens   ones

break apart

number

ACE IT TIME!

	yes	no
Did you underline the question in the word problem?	○	○
Did you circle the numbers or number words?	○	○
Did you box the clue words that tell you what operation to use?	○	○
Did you use a picture to show your thinking?	○	○
Did you label your numbers and your picture?	○	○
Did you explain your thinking and use math vocabulary words in your explanation?	○	○

Math on the Move

Count items around your home. Great items to count include coins, raisins in a snack box, or candy in a package. After counting, create groups of tens and ones to show the place value of the total number.

Addition Using the Arrow Method

FOLLOWING THE OBJECTIVE
You will add numbers using place value.

LEARN IT: You can add numbers by using the place value (100s, 10s, and 1s) of the second addend. Remember, addends are the two numbers you are adding together. An easy way to add larger numbers is to break down the second addend into tens and add the parts.

Example: 72 + 26 =

Find the sum using the arrow method.

Begin with the first addend and add parts of the second addend along arrows until you reach the second addend. Use numbers you can easily add like 100s, 10s, 5s, 2s, and 1s. Add the numbers along the arrows to find the total.

$$\underset{72}{} \xrightarrow{+10} \underset{82}{} \xrightarrow{+10} \underset{92}{} \xrightarrow{+2} \underset{94}{} \xrightarrow{+2} \underset{96}{} \xrightarrow{+2 \ (=26)} \underset{98}{}$$

There are several different ways you can add in parts. Select the values that are the easiest for you to add.

$$\underset{72}{} \xrightarrow{+20} \underset{92}{} \xrightarrow{+6 \ (=26)} \underset{98}{} \quad \text{OR} \quad \underset{72}{} \xrightarrow{+10} \underset{82}{} \xrightarrow{+10} \underset{92}{} \xrightarrow{+6 \ (=26)} \underset{98}{}$$

PRACTICE: Now you try

Find the sum using the arrow method.

1. 54 + 31 = _____ 54 ⟶	**2.** 24 + 45 = _____ 24 ⟶
3. 73 + 23 = _____ 73 ⟶	**4.** 62 + 13 = _____ 62 ⟶
5. 38 + 21 = _____ 38 ⟶	**6.** 85 + 22 = _____ 85 ⟶

7. 91 + 12 = _____

91 ⟶

8. 37 + 21 = _____

37 ⟶

9. 52 + 35 = _____

52 ⟶

10. 46 + 21 = _____

46 ⟶

The chart shows the number of car wash tickets sold by three players on the Panther's football team. Which two players sold a total of 69 tickets? Show your work and explain your thinking on a piece of paper.

Car Wash Ticket Sales	
Player	**Ticket Sales**
Michael	34
Martin	33
Micah	35

Math Vocabulary

add

sum

tens

ones

number

ACE IT TIME!

	yes	no
Did you underline the question in the word problem?	◯	◯
Did you circle the numbers or number words?	◯	◯
Did you box the clue words that tell you what operation to use?	◯	◯
Did you use a picture to show your thinking?	◯	◯
Did you label your numbers and your picture?	◯	◯
Did you explain your thinking and use math vocabulary words in your explanation?	◯	◯

Math on the Move

Play a game of "Guess My Number" with a friend or an adult. Think of any number between 10 and 100. Then add or subtract 10 in your head and tell your partner your new number. Can your partner guess your original number? *Hint:* your partner will have to add and subtract by 10 to find it!

Make a Ten and Add On

FOLLOWING THE OBJECTIVE
You will add within 20 using the "Make a Ten" strategy and know from memory all sums of two one-digit numbers.

LEARN IT: The "Make a Ten" strategy is a way to help you add facts with numbers near ten. First think, "How many more are needed to make 10?" Then ask, "How many are left over?" *Make the ten* and then add on the leftovers to find the sum.

Example: 8 + 7 =

How many more are needed to make 10?

```
8 + 7
|   /\
8  2 + 5
 \ /
10 + 5 = 15
```

> **think!**
> 8 is close to 10 ... I only need 2 more to make the 10! I can "take 2 out" from the 7 ...

If the 2 is taken from the 7, how many are left over? **5**

So, 8 + 7 can be made by making a 10 and then adding 5 to equal 15.

PRACTICE: Now you try

Show how you can make a ten to find the sum. Write the sum.

1. 8 + 5 = _____ /\ 2 3 10 + _____ = _____	2. 7 + 6 = _____ 10 + _____ = _____
3. 9 + 2 = _____ 10 + _____ = _____	4. 6 + 5 = _____ 10 + _____ = _____
5. 8 + 9 = _____ 10 + _____ = _____	6. 8 + 6 = _____ 10 + _____ = _____
7. 3 + 9 = _____ 10 + _____ = _____	8. 8 + 3 = _____ 10 + _____ = _____
9. 7 + 4 = _____ 10 + _____ = _____	10. 5 + 9 = _____ 10 + _____ = _____

Alex is thinking of a doubles fact. It has a sum that is greater than the sum of 6 + 7 but less than the sum of 6 + 9. What fact is Alex thinking of?

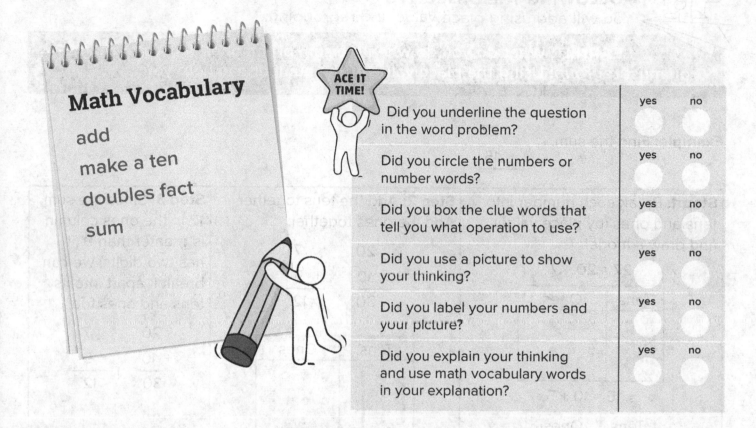

Math Vocabulary

add

make a ten

doubles fact

sum

ACE IT TIME!

	yes	no
Did you underline the question in the word problem?	◯	◯
Did you circle the numbers or number words?	◯	◯
Did you box the clue words that tell you what operation to use?	◯	◯
Did you use a picture to show your thinking?	◯	◯
Did you label your numbers and your picture?	◯	◯
Did you explain your thinking and use math vocabulary words in your explanation?	◯	◯

Math on the Move

Coins such as pennies, nickels, and dimes are helpful tools to use when practicing the skill of making ten and adding on. For each problem in this lesson, make each amount with change. For example, for 8 + 5, take 8 cents then add 5 more cents, group 10 cents together as 1 dime, and then count the total.

Regroup Using Place Value

FOLLOWING THE OBJECTIVE
You will add using place value and regrouping.

LEARN IT: What do you do when the sum of the ones is greater than 9? You must *regroup*! Using the place value of the number can help you.

Example: Find the sum.

$$27 + 15$$

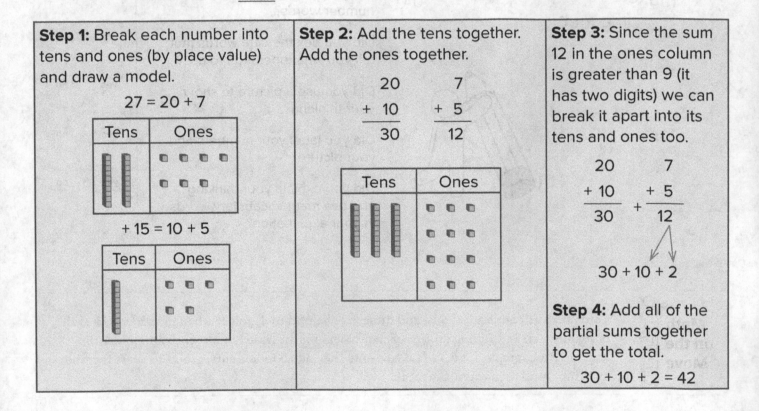

Step 1: Break each number into tens and ones (by place value) and draw a model.

$$27 = 20 + 7$$

Tens	Ones

$$+ 15 = 10 + 5$$

Tens	Ones

Step 2: Add the tens together. Add the ones together.

$$\begin{array}{r} 20 \\ + 10 \\ \hline 30 \end{array} \qquad \begin{array}{r} 7 \\ + 5 \\ \hline 12 \end{array}$$

Tens	Ones

Step 3: Since the sum 12 in the ones column is greater than 9 (it has two digits) we can break it apart into its tens and ones too.

$$\begin{array}{r} 20 \\ + 10 \\ \hline 30 \end{array} + \begin{array}{r} 7 \\ + 5 \\ \hline 12 \end{array}$$

$$30 + 10 + 2$$

Step 4: Add all of the partial sums together to get the total.

$$30 + 10 + 2 = 42$$

PRACTICE: Now you try

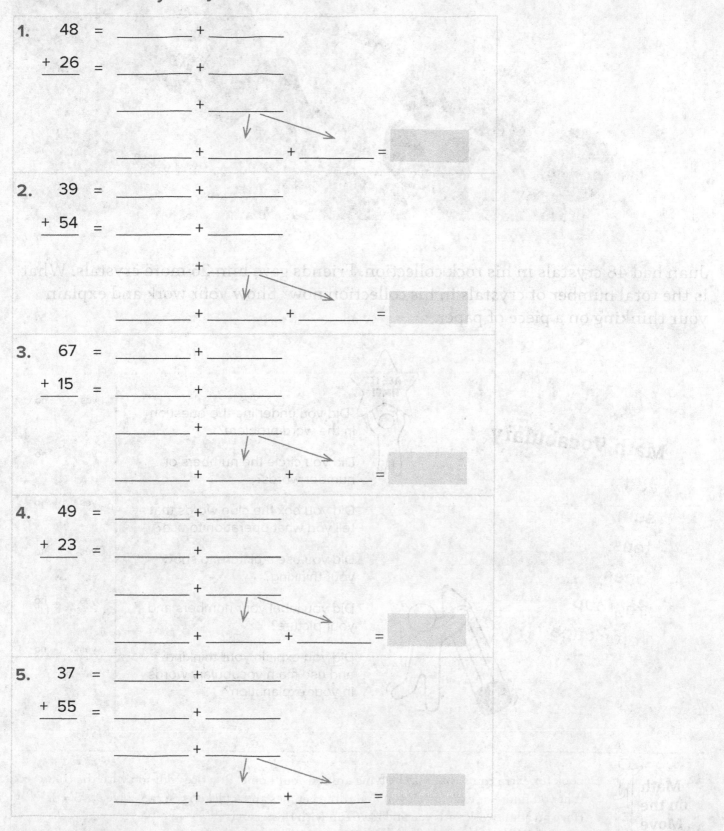

1. 48 = _____ + _____

    + 26 = _____ + _____

    _____ + _____

    _____ + _____ + _____ =

2. 39 = _____ + _____

    + 54 = _____ + _____

    _____ + _____ + _____ =

3. 67 = _____ + _____

    + 15 = _____ + _____

    _____ + _____

    _____ + _____ + _____ =

4. 49 = _____ + _____

    + 23 = _____ + _____

    _____ + _____

    _____ + _____ + _____ =

5. 37 = _____ + _____

    + 55 = _____ + _____

    _____ + _____

    _____ + _____ + _____ =

Juan had 48 crystals in his rock collection. Friends gave him 26 more crystals. What is the total number of crystals in his collection now? Show your work and explain your thinking on a piece of paper.

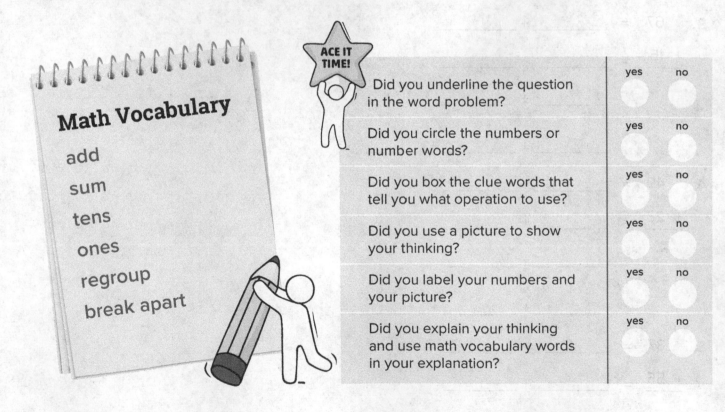

Math Vocabulary

add

sum

tens

ones

regroup

break apart

ACE IT TIME!

	yes	no
Did you underline the question in the word problem?	○	○
Did you circle the numbers or number words?	○	○
Did you box the clue words that tell you what operation to use?	○	○
Did you use a picture to show your thinking?	○	○
Did you label your numbers and your picture?	○	○
Did you explain your thinking and use math vocabulary words in your explanation?	○	○

Math on the Move

Look for extra coins that may be lying around your home. Practice addition with the coins that you find. For example, if you find three pennies and a nickel, you can solve 3 + 5 = 8. Have an adult work with you and have fun with this exercise!

Add Numbers in a Series

FOLLOWING THE OBJECTIVE
You will add up to four two-digit numbers in a series.

LEARN IT: If you can add two numbers, is it just as easy to add three numbers? What about more than three? There are different strategies you can use to add three or more numbers.

Example: Consider this problem: Mark has 32 red marbles, 14 blue marbles, and 46 green marbles. How many marbles does Mark have in all?

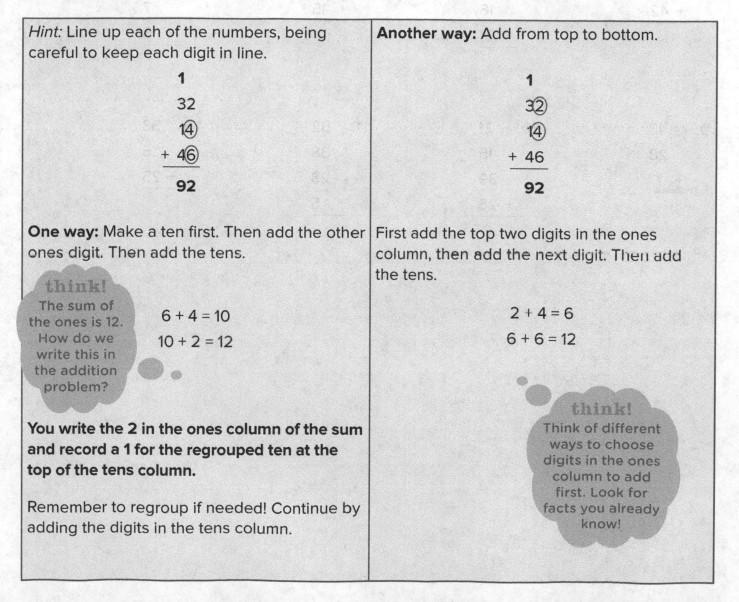

Hint: Line up each of the numbers, being careful to keep each digit in line.

$$1$$
$$32$$
$$14$$
$$+\ 46$$
$$92$$

One way: Make a ten first. Then add the other ones digit. Then add the tens.

think! The sum of the ones is 12. How do we write this in the addition problem?

$6 + 4 = 10$
$10 + 2 = 12$

You write the 2 in the ones column of the sum and record a 1 for the regrouped ten at the top of the tens column.

Remember to regroup if needed! Continue by adding the digits in the tens column.

Another way: Add from top to bottom.

$$1$$
$$32$$
$$14$$
$$+\ 46$$
$$92$$

First add the top two digits in the ones column, then add the next digit. Then add the tens.

$2 + 4 = 6$
$6 + 6 = 12$

think! Think of different ways to choose digits in the ones column to add first. Look for facts you already know!

Unit 4: Addition and Subtraction Concepts

PRACTICE: Now you try

1.
```
   13
   28
   15
 + 22
```

2.
```
   46
   19
 + 24
```

3.
```
   60
   16
 + 12
```

4.
```
   23
   42
   16
 + 48
```

5.
```
   31
   18
 + 42
```

6.
```
   13
   14
   16
 + 21
```

7.
```
   85
   38
 + 15
```

8.
```
   60
   13
    7
 +  5
```

9.
```
   43
   28
 + 17
```

10.
```
   21
   15
   39
 + 25
```

11.
```
   52
   38
   23
 +  5
```

12.
```
   33
   15
 + 25
```

Look at the numbers in the chart. Find three numbers that have a sum of 97. Here are two clues to help you:

16	14	17
48	25	35

➡ One number is the sum of 7 and 7.

➡ One number is between 30 and 40.

Show your work and explain your thinking on a piece of paper.

Math Vocabulary

add

doubles fact

ones

regroup

ACE IT TIME!

	yes	no
Did you underline the question in the word problem?	○	○
Did you circle the numbers or number words?	○	○
Did you box the clue words that tell you what operation to use?	○	○
Did you use a picture to show your thinking?	○	○
Did you label your numbers and your picture?	○	○
Did you explain your thinking and use math vocabulary words in your explanation?	○	○

Math on the Move

Look in a newspaper or mail flyer with advertisements. Select three to four items you would like to have and then add to find the total cost of your items. Happy shopping!

Unit 4: Addition and Subtraction Concepts

Add Three-Digit Numbers

FOLLOWING THE OBJECTIVE
You will add three-digit numbers within 1,000.

LEARN IT: When you add three-digit numbers, sometimes you will regroup more than once. Remember: if the sum of the ones or the tens is greater than 9, you will need to regroup!

think!
Are there 10 or more ones? Are there 10 or more tens?

Example: Jayla had 336 shells in her collection. Cameron had 198 shells. How many shells did they have in all?

Step 1: Line up each of the numbers, being careful to keep each digit in line. Add the ones digits.

$$\begin{array}{r} 1 \\ 336 \\ + 198 \\ \hline 4 \end{array}$$

6 ones + 8 ones = 14 ones

think!
The sum of the ones is 14. How do we write this in the addition problem?
You write the 4 in the ones column of the sum and record a 1 for the regrouped ten at the top of the tens column.

Step 2: Add the digits in the tens column.

$$\begin{array}{r} 11 \\ 336 \\ + 198 \\ \hline 34 \end{array}$$

1 ten + 3 tens + 9 tens = 13 tens

think!
The sum of the tens is 13. How do we write this in the addition problem?
You write the 3 in the tens column of the sum and record a 1 for the regrouped ten at the top of the hundreds column.

Step 3: Add the digits in the hundreds column.

$$\begin{array}{r} 11 \\ 336 \\ + 198 \\ \hline 534 \end{array}$$

1 hundred + 3 hundreds + 1 hundred = 5 hundreds

PRACTICE: Now you try

1. 624 + 310	2. 332 + 164	3. 548 + 263	4. 452 + 249
5. 293 + 327	6. 426 + 546	7. 531 + 175	8. 109 + 346
9. 184 + 278	10. 546 + 329	11. 623 + 272	12. 234 + 457

Juan, Julio, and Juanita were keeping a record of the minutes they spent reading at home last month. Which two students read a total of 600 minutes? Show your work and explain your thinking on a piece of paper.

Student	Minutes Read
Juan	327
Julio	237
Juanita	273

Math Vocabulary

add

break apart

sum

tens

ones

regroup

ACE IT TIME!

	yes	no
Did you underline the question in the word problem?		
Did you circle the numbers or number words?		
Did you box the clue words that tell you what operation to use?		
Did you use a picture to show your thinking?		
Did you label your numbers and your picture?		
Did you explain your thinking and use math vocabulary words in your explanation?		

Math on the Move

Find two of your favorite books that you have read and see how many pages they each have. Add them together and celebrate the fact that you have read so many pages!

Subtraction Using Place Value

FOLLOWING THE OBJECTIVE
You will subtract numbers using place value.

LEARN IT: You now know that numbers can be broken apart by using their place value (100s, 10s, and 1s). You can use place value to subtract larger numbers, just like you did with addition!

Example: Find the difference. *Hint:* Finding the difference means we are subtracting!

$$\begin{array}{r} 64 \\ -\ 53 \\ \hline \end{array}$$

Step 1: Break each number into tens and ones.	**Step 2:** Subtract the tens. Subtract the ones.	**Step 3:** Add the two differences together to get the final answer.
$64 = 60 + 4$ $-53 = -50 + 3$	$\begin{array}{r}60\\-50\\\hline 10\end{array}$ $\begin{array}{r}4\\-3\\\hline 1\end{array}$	$10 + 1 = 11$ So, $64 - 53 = 11$.

PRACTICE: Now you try Break each number into tens and ones. Solve.

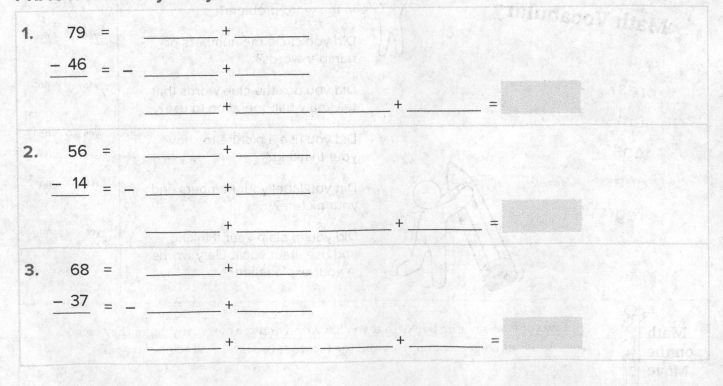

1. $79 =$ _____ + _____

    $-\ 46 =\ -$ _____ + _____

    _____ + _____ _____ + _____ = _____

2. $56 =$ _____ + _____

    $-\ 14 =\ -$ _____ + _____

    _____ + _____ _____ + _____ = _____

3. $68 =$ _____ + _____

    $-\ 37 =\ -$ _____ + _____

    _____ + _____ _____ + _____ = _____

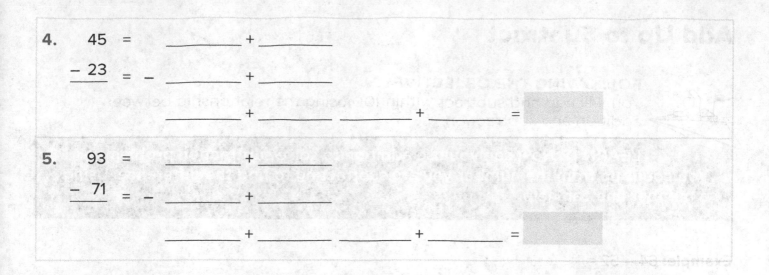

4. 45 = _____ + _____

  − 23 = − _____ + _____

  _____ + _____ _____ + _____ = [   ]

5. 93 = _____ + _____

  − 71 = − _____ + _____

  _____ + _____ _____ + _____ = [   ]

An after-school recreation program had 57 toys in a box. Some of the toys were taken out of the box during game time. Now there are 34 toys in the box. How many toys were taken out of the box? Show your work and explain your thinking on a piece of paper.

Math Vocabulary

subtract

difference

tens

ones

break apart

number

ACE IT TIME!

	yes	no
Did you underline the question in the word problem?	○	○
Did you circle the numbers or number words?	○	○
Did you box the clue words that tell you what operation to use?	○	○
Did you use a picture to show your thinking?	○	○
Did you label your numbers and your picture?	○	○
Did you explain your thinking and use math vocabulary words in your explanation?	○	○

Math on the Move

Practice subtraction with a fun game of "Count Down." You will need two number cubes, a sheet of paper, and a pencil to play. Each player writes the number 99 at the top of the piece of paper. You roll both number cubes and add up the numbers you rolled. Subtract that sum from 99. The first player to get to zero wins!

Add Up to Subtract

FOLLOWING THE OBJECTIVE
You will add and subtract within 100 using the relationship between addition and subtraction.

LEARN IT: How can you use addition to solve subtraction problems? You count up from the number you are subtracting to find the difference. Using a number line can help.

Example: 64 − 52 =

Start at 52. Count up to 62. That is 10. Now count up 2 more to 64.

> **think!**
> 10 + 2 = 12
> So 64 − 52 = 12

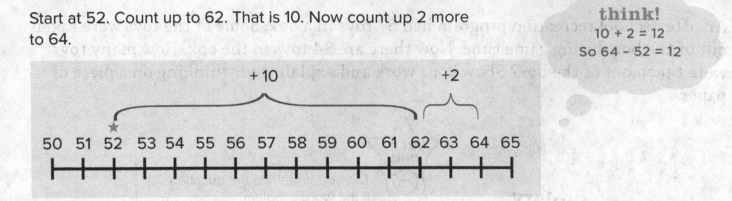

PRACTICE: Now you try

1. 53 − 45 = _____

40 41 42 43 44 45 46 47 48 49 50 51 52 53 54 55

2. 75 − 68 = _____

60 61 62 63 64 65 66 67 68 69 70 71 72 73 74 75

3. 34 − 29 = _____

20 21 22 23 24 25 26 27 28 29 30 31 32 33 34 35

4. 97 − 88 = _____

85 86 87 88 89 90 91 92 93 94 95 96 97 98 99 100

5. 79 – 63 = _____

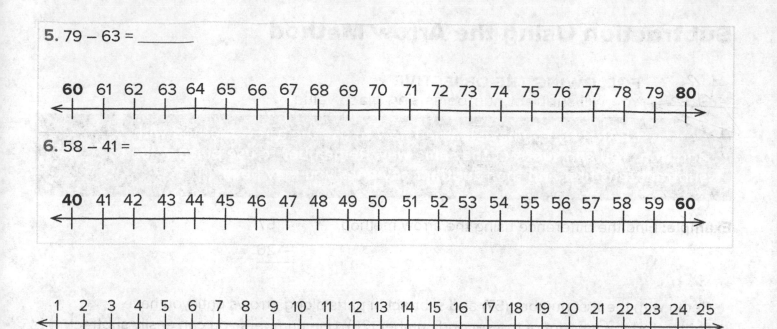

60 61 62 63 64 65 66 67 68 69 70 71 72 73 74 75 76 77 78 79 80

6. 58 – 41 = _____

40 41 42 43 44 45 46 47 48 49 50 51 52 53 54 55 56 57 58 59 60

1 2 3 4 5 6 7 8 9 10 11 12 13 14 15 16 17 18 19 20 21 22 23 24 25

Christine did 7 fewer math problems than Crystal did. Crystal solved 22 math problems. How many problems did Christine solve? Use the number line to help you solve. Show your work and explain your thinking on a piece of paper.

ACE IT TIME!

	yes	no
Did you underline the question in the word problem?	yes ○	no ○
Did you circle the numbers or number words?	yes ○	no ○
Did you box the clue words that tell you what operation to use?	yes ○	no ○
Did you use a picture to show your thinking?	yes ○	no ○
Did you label your numbers and your picture?	yes ○	no ○
Did you explain your thinking and use math vocabulary words in your explanation?	yes ○	no ○

Math on the Move

There are several ways you can count up in each problem. Choose a problem on this page and show a different way to use addition to find the difference. See how many different ways you could add up to find the difference.

Subtraction Using the Arrow Method

FOLLOWING THE OBJECTIVE
You will subtract numbers using place value.

LEARN IT: You have learned how to use the arrow method with addition. You can subtract the same way. Remember to use place value to subtract tens first and then subtract any ones you may have.

Example: Find the difference using the arrow method.

$$57$$
$$- 26$$

Begin with the first number (57) and subtract in parts along arrows until you have subtracted the amount of the second number (26). Use numbers you can easily subtract like 100s, 10s, 5s, 2s, and 1s. Subtract the numbers along the arrows to find the difference.

$$
\begin{array}{ccccccccc}
& -10 & & -10 & & -2 & & -2 & & -2 \ (=26) \\
57 & \longrightarrow & 47 & \longrightarrow & 37 & \longrightarrow & 35 & \longrightarrow & 33 & \longrightarrow & 31
\end{array}
$$

There are several different ways you can subtract in parts. Select the values that are the easiest for you to subtract.

$$
\begin{array}{ccccc}
& -20 & & -6 \ (=26) \\
57 & \longrightarrow & 37 & \longrightarrow & 31
\end{array}
\quad \text{OR} \quad
\begin{array}{ccccccc}
& -10 & & -10 & & -6 \ (=26) \\
57 & \longrightarrow & 47 & \longrightarrow & 37 & \longrightarrow & 31
\end{array}
$$

PRACTICE: Now you try

Find the difference using the arrow method.

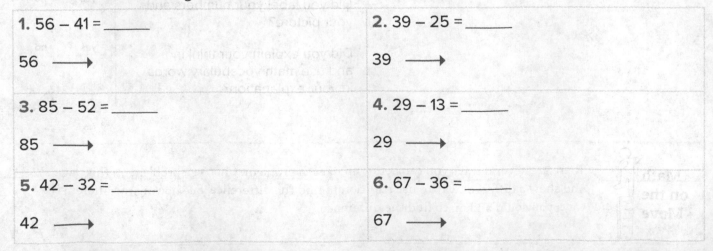

1. 56 – 41 = _____ 56 ⟶	**2.** 39 – 25 = _____ 39 ⟶
3. 85 – 52 = _____ 85 ⟶	**4.** 29 – 13 = _____ 29 ⟶
5. 42 – 32 = _____ 42 ⟶	**6.** 67 – 36 = _____ 67 ⟶

7. 95 – 61 = _____

95 ⟶

8. 56 – 21 = _____

56 ⟶

9. 59 – 17 = _____

59 ⟶

10. 66 – 45 = _____

66 ⟶

Mrs. Blake had 96 math papers to grade. She graded 35 papers on Monday and 35 more papers on Tuesday. How many papers will she have left to grade on Wednesday? Show your work and explain your thinking on a piece of paper.

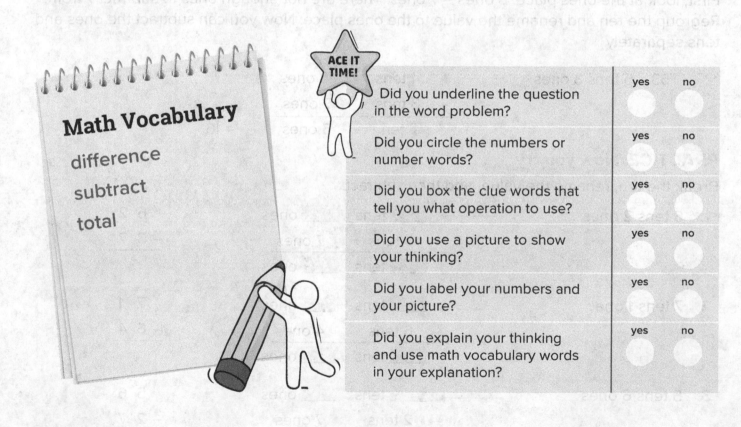

Math Vocabulary

difference

subtract

total

ACE IT TIME!

	yes	no
Did you underline the question in the word problem?	○	○
Did you circle the numbers or number words?	○	○
Did you box the clue words that tell you what operation to use?	○	○
Did you use a picture to show your thinking?	○	○
Did you label your numbers and your picture?	○	○
Did you explain your thinking and use math vocabulary words in your explanation?	○	○

Math on the Move

Remember, subtraction is finding the difference! Look for ways we use subtraction in our daily life. For example, find a book you or someone at home is reading. How can you figure out how many pages there are left to read?

Subtract Two-Digit Numbers with Regrouping

FOLLOWING THE OBJECTIVE
You will subtract two-digit numbers within 100.

LEARN IT: When you are subtracting, it is important to start in the ones place, just like with adding!

Example:

$$53 = 5 \text{ tens } 3 \text{ ones}$$
$$- 37 = 3 \text{ tens } 7 \text{ ones}$$
$$16$$

> **think!**
> You are renaming 5 tens and 3 ones as 4 tens and 13 ones! Can you explain why the number in the tens place only goes down by 1? How do 5 tens become 4 tens?

First, look at the ones place: 3 ones – 7 ones. There are not enough ones to subtract 7 from. Regroup the ten and rename the value to the ones place. Now you can subtract the ones and tens separately.

$$53 = 5 \text{ tens } 3 \text{ ones} \longrightarrow$$

	4 tens	13 ones	
–	3 tens	7 ones	
	1 ten	6 ones	= 16

PRACTICE: Now you try

Break the ten, rename the value, and then subtract.

1. 6 tens 2 ones ⟶

	▢ tens	▢ ones		6 2
–	3 tens	7 ones		– 3 7
	▢ tens	▢ ones		

2. 7 tens 1 one ⟶

	▢ tens	▢ ones		7 1
–	6 tens	4 ones		– 6 4
	▢ tens	▢ ones		

3. 5 tens 6 ones ⟶

	▢ tens	▢ ones		5 6
–	2 tens	7 ones		– 2 7
	▢ tens	▢ ones		

4. 6 tens 3 ones ⟶

	▢ tens	▢ ones		6 3
–	4 tens	5 ones		– 4 5
	▢ tens	▢ ones		

PRACTICE: Now you try Solve. Write the difference.

5.	6.	7.	8.	9.
32 − 16	70 − 38	85 − 47	64 − 56	55 − 9

10.	11.	12.	13.	14.
20 − 12	52 − 17	43 − 28	47 − 8	66 − 48

Allison's family toured a local art museum. They noticed there were 24 paintings done by French artists. They saw 51 paintings by American artists. They also saw 16 glass sculptures and 37 photographs. How many more paintings than photographs did they see? Show your work and explain your thinking on a piece of paper.

Math Vocabulary

total

difference

regroup

tens

ones

ACE IT TIME!

	yes	no
Did you underline the question in the word problem?	○	○
Did you circle the numbers or number words?	○	○
Did you box the clue words that tell you what operation to use?	○	○
Did you use a picture to show your thinking?	○	○
Did you label your numbers and your picture?	○	○
Did you explain your thinking and use math vocabulary words in your explanation?	○	○

Math on the Move A newspaper reports that a team has beaten another team by six points. How did they get that number? They subtracted one team's score from the other. Find the scores of sports teams in local newspapers or online and subtract them.

Subtract Three-Digit Numbers with Regrouping

FOLLOWING THE OBJECTIVE
You will subtract three-digit numbers within 1,000.

LEARN IT: Subtracting a three-digit number is similar to subtracting a two-digit number. Remember to ask yourself, *"Do I need to regroup?"* for each place value: the ones, the tens, and the hundreds.

think!
Remember, you are renaming the numbers based on their new value! So after you regroup, 563 is equal to 4 hundreds, 15 tens, and 13 ones!

Example:

$$
\begin{array}{r}
563 \\
-\ 285 \\
\end{array}
$$

Write the numbers vertically. Line up each number by place value. Subtract the ones, then the tens, and then the hundreds.

Subtract ones	Subtract tens	Subtract hundreds
3 < 5 so it's time to regroup!	5 < 8 so regroup	4 > 2 so subtract
5 13	4 15 13	4 15 13
5 6̸ 3	5̸ 6̸ 3	5̸ 6̸ 3
− 2 8 **5**	− 2 **8** 5	− **2** 8 5
8	**7** 8	**2** 7 8

PRACTICE: Now you try

Solve. Write the difference.

1.	2.	3.	4.	5.	6.
425 − 148	268 − 193	634 − 367	756 − 273	589 − 67	368 − 271

7.	8.	9.	10.	11.	12.
342 − 138	463 − 182	856 − 497	381 − 146	537 − 281	673 − 574

PRACTICE: Now you try

Write the missing digits.

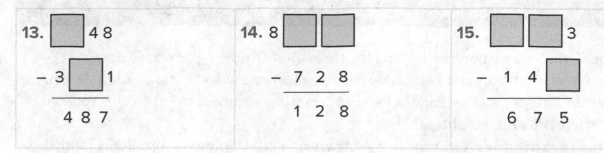

13.
```
  ☐ 4 8
- 3 ☐ 1
_____
  4 8 7
```

14.
```
  8 ☐ ☐
- 7 2 8
_____
  1 2 8
```

15.
```
  ☐ ☐ 3
- 1 4 ☐
_____
  6 7 5
```

Solve this number riddle: I am a three-digit number. If you subtract me from 500, you are left with 247. What number am I? Show your work and explain your thinking on a piece of paper.

Math Vocabulary

subtract

difference

regroup

hundreds

tens

ones

ACE IT TIME!

	yes	no
Did you underline the question in the word problem?	○	○
Did you circle the numbers or number words?	○	○
Did you box the clue words that tell you what operation to use?	○	○
Did you use a picture to show your thinking?	○	○
Did you label your numbers and your picture?	○	○
Did you explain your thinking and use math vocabulary words in your explanation?	○	○

Math on the Move Use addition to check each of the problems in this section. To prove your answer is correct, add it to the bottom number in each problem. If it equals the top number—you are right!

REVIEW

Congratulations! You've finished the lessons for Units 3–4. This means you've noticed the relationship between addition and subtraction. You've learned ways to make addition and subtraction easier. You've practiced adding and subtracting numbers up to 1,000. You can even regroup when you add or subtract.

Now it's time to prove your addition and subtraction skills. Solve the problems below! Use all of the methods you have learned.

Activity Section 1

Write the related facts for each fact family.

1. 6 + 3 = _____

2. 7 – 4 = _____

3. 2 + 8 = _____

4. 12 – 9 = _____

5. 7 + 8 = _____

6. 14 – 6 = _____

Solve each problem. Watch the operation signs.

7.	8	8.	6	9.	10	10.	5	11.	12	12.	1	13.	9
	+ 4		– 3		– 5		+ 7		– 6		+ 5		– 9

14.	3	15.	9	16.	11	17.	6	18.	9	19.	6	20.	17
	+ 2		– 4		– 9		+ 0		– 7		+ 4		– 8

21.	12	22.	2	23.	8	24.	10	25.	16	26.	7	27.	1
	– 2		+ 1		+ 2		– 7		– 8		+ 7		+ 9

Activity Section 2

Solve each problem. Watch the operation signs.

1. $\begin{array}{r} 32 \\ + 43 \\ \hline \end{array}$	2. $\begin{array}{r} 19 \\ + 12 \\ \hline \end{array}$	3. $\begin{array}{r} 549 \\ + 374 \\ \hline \end{array}$	4. $\begin{array}{r} 149 \\ + 475 \\ \hline \end{array}$	5. $\begin{array}{r} 31 \\ 17 \\ 24 \\ + 45 \\ \hline \end{array}$
6. $\begin{array}{r} 37 \\ + 48 \\ \hline \end{array}$	7. $\begin{array}{r} 60 \\ 14 \\ + 22 \\ \hline \end{array}$	8. $\begin{array}{r} 822 \\ + 134 \\ \hline \end{array}$	9. $\begin{array}{r} 107 \\ + 572 \\ \hline \end{array}$	10. $\begin{array}{r} 189 \\ + 543 \\ \hline \end{array}$

Activity Section 3

Solve each problem. Write the difference.

1. $\begin{array}{r} 18 \\ - 16 \\ \hline \end{array}$	2. $\begin{array}{r} 17 \\ - 11 \\ \hline \end{array}$	3. $\begin{array}{r} 15 \\ - 10 \\ \hline \end{array}$	4. $\begin{array}{r} 13 \\ - 12 \\ \hline \end{array}$	5. $\begin{array}{r} 19 \\ - 15 \\ \hline \end{array}$
6. $\begin{array}{r} 16 \\ - 16 \\ \hline \end{array}$	7. $\begin{array}{r} 14 \\ - 10 \\ \hline \end{array}$	8. $\begin{array}{r} 20 \\ - 10 \\ \hline \end{array}$	9. $\begin{array}{r} 16 \\ - 15 \\ \hline \end{array}$	10. $\begin{array}{r} 15 \\ - 13 \\ \hline \end{array}$

Solve each problem. Regroup if necessary. Write the difference.

11. $\begin{array}{r} 35 \\ - 17 \\ \hline \end{array}$	12. $\begin{array}{r} 59 \\ - 29 \\ \hline \end{array}$	13. $\begin{array}{r} 80 \\ - 47 \\ \hline \end{array}$	14. $\begin{array}{r} 72 \\ - 58 \\ \hline \end{array}$	15. $\begin{array}{r} 27 \\ - 13 \\ \hline \end{array}$
16. $\begin{array}{r} 507 \\ -294 \\ \hline \end{array}$	17. $\begin{array}{r} 483 \\ - 127 \\ \hline \end{array}$	18. $\begin{array}{r} 837 \\ -324 \\ \hline \end{array}$	19. $\begin{array}{r} 378 \\ -259 \\ \hline \end{array}$	20. $\begin{array}{r} 611 \\ - 543 \\ \hline \end{array}$

Activity Section 4

Use a problem-solving strategy to help you.

1. Kaitlyn had 938 pennies. She took 900 pennies to the bank. How many pennies does Kaitlyn have left?

2. Jennifer's family took a trip. On the first day, they drove 256 miles. On the second day, they drove 391 miles. How many miles did her family drive in all?

UNDERSTAND

Understand the meaning of what you have learned and apply your knowledge.

You can use place value understanding to add numbers within 100. You have learned there are different ways to add and subtract. Let's look at some other people's work below.

Activity Section

Four students in Mrs. Colson's class used different strategies to show how they solved the problem 62 + 28.

Chloe used the Arrow method: $62 \xrightarrow{+20} 72 \xrightarrow{+8} 80$	Charley used the Place-Value method: $62 = 60 + 2$ $+\ \ 28 = 20 + 8$ $\overline{\qquad\ \ 80 + 10}$ $= 90$
Kyle made a ten and then added on: $\begin{array}{ccc} 62 & + & 28 \\ +8 & & +2 \\ \hline 70 & + & 30 \\ & = 100 & \end{array}$	Ginger used the Standard method: $\begin{array}{r} 0 \\ 62 \\ +\ \ 28 \\ \hline 81 \end{array}$

After checking their answers, Mrs. Colson discovered only one student solved the problem correctly.

1. Find the student who solved the problem correctly.

2. Explain what each of the other three students did wrong.

Chloe	Charley
Kyle	Ginger

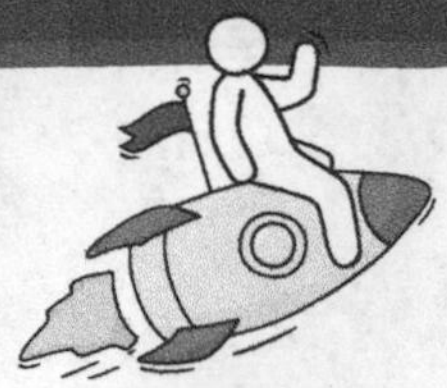

DISCOVER

Discover how you can apply the information you have learned.

Adding and subtracting is a skill we use daily. Look for patterns within the hundreds chart to help you.

Activity Section

This is a hundreds chart. It is missing many numbers! Use the number clues on the chart to identify the value of the dark-gray squares.

11									
				25					
	42								
						57			
		63							
								79	
							88		

How did you choose the numbers you did for each dark-gray square?

____

____

____

____

____

____

Money and Time Concepts

What Time Is It?

FOLLOWING THE OBJECTIVE
You will tell and write time to the nearest five minutes.

LEARN IT: You can use an analog or a digital clock to tell time. Remember, there are 60 minutes in one hour!

Let's look at this **analog clock**. The numbers 1–12 on the inside represent the hours. The numbers on the outside are counting by fives and represent the minutes.

Now let's look at the hands on the clock. **The short hand is the hour hand.** It takes 60 minutes (1 hour) for the short hand to move from one number to the next. When the short hand is between two numbers, that means it is still moving toward the next hour. For example, the short hand on this clock is pointed between 1 and 2. It is not yet 2 o'clock!

The long hand is the minute hand. It takes one minute for the long hand to move from one tick mark to the next. It is easy to count by 5s in minutes because there are 5 tick marks between each minute.

How many minutes does this clock show? Start at the top of the clock (at the number 12) and skip count by 5s until you reach the big hand ... *5, 10, 15, 20, 25, 30, 35, 40, 45, 50!*

You can also tell this time on a **digital clock**. Notice that the first number tells us the hour, and the second number tells us the minute. What time does this clock say?

1:50

HOUR:MINUTE

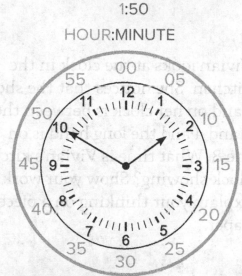

PRACTICE: Now you try Write the time shown on each clock.

1.

2.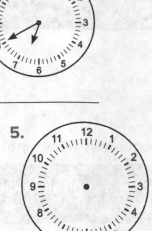

3. **3:25**

4. **9:50**

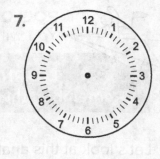

_____   _____   _____   _____

Look at the time on the line below each clock. Draw the long and short hands on the clock to show the correct time.

5.   6.   7.

7:10   10:35   4:45

Vivian looks at the clock in the kitchen. She notices that the short hand on her clock is between the 3 and 4 and the long hand is on the 8. What time is Vivian's kitchen clock showing? Show your work and explain your thinking on a piece of paper.

ACE IT TIME!

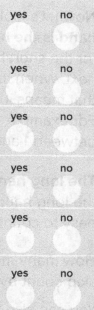

	yes	no
Did you underline the question in the word problem?	yes	no
Did you circle the numbers or number words?	yes	no
Did you box the clue words that tell you what operation to use?	yes	no
Did you use a picture to show your thinking?	yes	no
Did you label your numbers and your picture?	yes	no
Did you explain your thinking and use math vocabulary words in your explanation?	yes	no

Math Vocabulary

hour hand

minute hand

analog clock

Math on the Move Practice telling and writing time from an analog clock at home. If there is no analog clock at home, then you can draw one and label it. Use it to practice telling and writing time.

A.M. or P.M.?

FOLLOWING THE OBJECTIVE
You will tell and write time to the nearest five minutes using A.M. and P.M.

LEARN IT: By now you have noticed that there are only twelve hours showing on a clock, but there are 24 hours in a day. So how can you tell the difference between daytime and nighttime hours? Midnight is 12 *A.M.* Noon is 12 *P.M.* The time after midnight and before noon is called A.M. The time after noon and before midnight is called P.M.

At 8:00 A.M. you might be eating breakfast.

8:00

At 8:00 P.M. you might be in bed sleeping.

8:00

PRACTICE: Now you try Write the time on the digital clock. Circle A.M. or P.M.

1. Doing homework

   A.M.   P.M.

2. Eating lunch

   A.M.   P.M.

3. Eating dinner

   A.M.   P.M.

4. Waking up

   A.M.   P.M.

Use the time on the digital clock to draw in the short and long hands. Circle A.M. or P.M.

5. Going to math class

A.M.   P.M.

1 : 45

6. Watching the sunset

A.M.   P.M.

8 : 30

Julio says that at 2:00 A.M. he is at school. Matthew says that at 2:00 A.M. he is sleeping. Who is correct? Explain. Show your work and explain your thinking on a piece of paper.

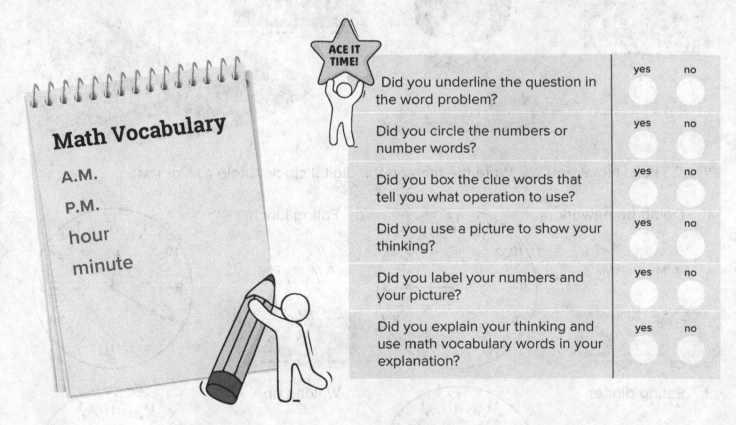

Math Vocabulary

A.M.

P.M.

hour

minute

ACE IT TIME!

	yes	no
Did you underline the question in the word problem?	yes	no
Did you circle the numbers or number words?	yes	no
Did you box the clue words that tell you what operation to use?	yes	no
Did you use a picture to show your thinking?	yes	no
Did you label your numbers and your picture?	yes	no
Did you explain your thinking and use math vocabulary words in your explanation?	yes	no

Math on the Move Make your own activity cards! Fold a piece of paper into six equal rectangles. You may need an adult to show you how. Cut on each folded line. On each card, draw a different picture of yourself doing an activity that you do every day. Ask a friend to guess whether the activity is done in the A.M. or in the P.M.

Dollars and Cents

FOLLOWING THE OBJECTIVE
You will solve problems using dollars, quarters, dimes, nickels, and pennies.

LEARN IT: You can find the value of a group of bills and coins. The chart below shows you the value of each coin. Use what you know about skip counting to help!

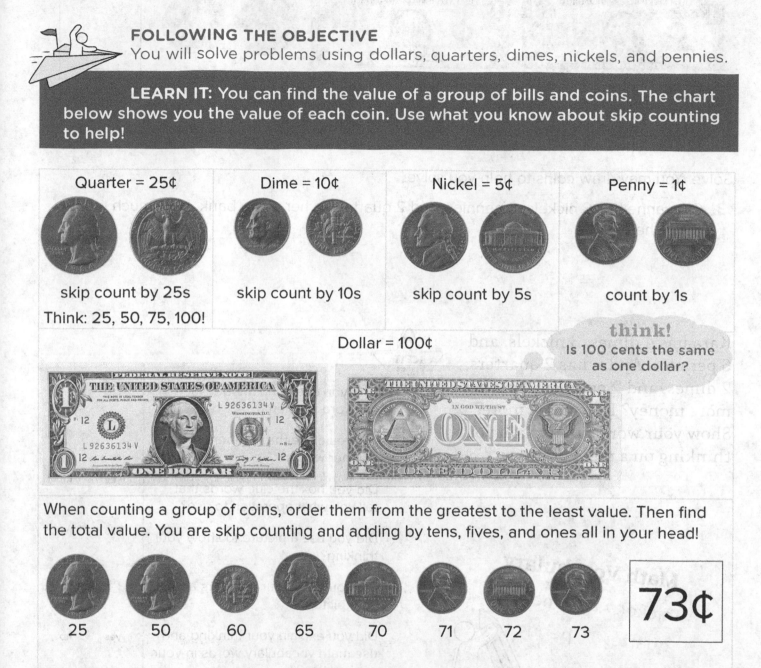

Quarter = 25¢　skip count by 25s　Think: 25, 50, 75, 100!

Dime = 10¢　skip count by 10s

Nickel = 5¢　skip count by 5s

Penny = 1¢　count by 1s

Dollar = 100¢

think! Is 100 cents the same as one dollar?

When counting a group of coins, order them from the greatest to the least value. Then find the total value. You are skip counting and adding by tens, fives, and ones all in your head!

25　50　60　65　70　71　72　73　**73¢**

PRACTICE: Now you try Write the total value of the money in the box.

1. Charlotte has these coins. How much money does she have?

2. Roger has these dollars and coins in his pocket. How much money does he have?

Solve. You may draw coins to help you solve.

3. Cheyanne has 6 nickels, 6 pennies, and 2 quarters in her piggy bank. How much money does she have?

Kara has 4 dimes, 3 nickels, and 6 pennies. Andre has 2 quarters, 2 dimes, and 2 pennies. Who has more money? How much more? Show your work and explain your thinking on a piece of paper.

ACE IT TIME!

	yes	no
Did you underline the question in the word problem?	○	○
Did you circle the numbers or number words?	○	○
Did you box the clue words that tell you what operation to use?	○	○
Did you use a picture to show your thinking?	○	○
Did you label your numbers and your picture?	○	○
Did you explain your thinking and use math vocabulary words in your explanation?	○	○

Math Vocabulary

nickel   dime

quarter   penny

cents

Math on the Move

Ask an adult if you can use any change lying around your home to help you practice counting different money amounts. Use this money to help you practice skip counting by fives, tens, and twenty-fives.

Make Money Different Ways

FOLLOWING THE OBJECTIVE
You will solve problems using dollars, quarters, dimes, nickels, and pennies.

LEARN IT: Can you think of all the ways you can add to make 10? There are many different ways to make 10, but the value equals 10 each time. Just like adding, you can show money amounts in different ways!

Example: Show 35¢ in two different ways.

Remember the value of each coin:

| Quarter = 25¢ | Dime = 10¢ | Nickel = 5¢ | Penny = 1¢ |

Now put the coins in groups from greatest value to least (quarters to pennies) and count up to find their value. Can you think of different ways?

25 + 10 = 35

35¢

10 + 10 + 10 + 1 + 1 + 1 + 1 + 1

think!
Are there any other ways to make 35 cents?

PRACTICE: Now you try

Find two ways to show the given amount.

1. 63¢		
2. 78¢		
3. 29¢		
4. 55¢		
5. 42¢		
6. 66¢		
7. 33¢		

Megan has 2 quarters, 3 dimes, 2 nickels, and 3 pennies. How much money does Megan have in all? In the space below, show how Megan can use different coins to make the same amount of money.

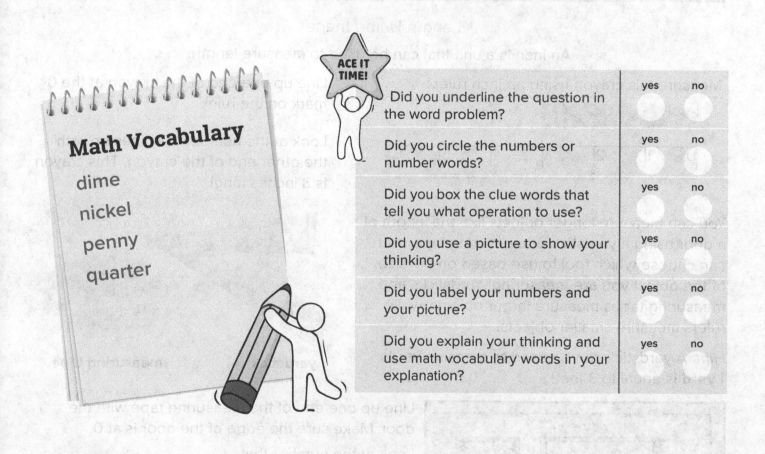

Math Vocabulary

dime

nickel

penny

quarter

ACE IT TIME!

	yes	no
Did you underline the question in the word problem?	yes	no
Did you circle the numbers or number words?	yes	no
Did you box the clue words that tell you what operation to use?	yes	no
Did you use a picture to show your thinking?	yes	no
Did you label your numbers and your picture?	yes	no
Did you explain your thinking and use math vocabulary words in your explanation?	yes	no

Math on the Move

Try to make different money amounts in as many ways as you can. Make 59¢ as many ways as you can. Record it by drawing pictures to show your different ways. Now try making 85¢ as many ways as you can. Have an adult check your drawings. You're on a roll now ... keep on going!

Measurement and Data Concepts

Measure the Length

FOLLOWING THE OBJECTIVE
You will practice measuring the length of objects using the appropriate tool.

LEARN IT: You can use different units to measure how long an object is. In this lesson, we will use inches or centimeters to measure length.

Length Using Inches

An inch is a unit that can be used to measure length.

Measure this crayon using an inch ruler.

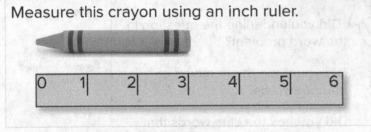

Line up one end of the crayon at the 0 mark on the ruler.

Look at the number that lines up with the other end of the crayon. This crayon is 3 inches long!

You can measure longer objects like the length of a door using a yardstick or measuring tape. You can choose which tool to use based on the size of the object you are measuring! Yardsticks and measuring tapes measure larger objects, while rulers measure smaller objects.

Hint: A yardstick can be used to measure yards. **1 yard** is equal to 3 feet!

yardstick   **measuring tape**

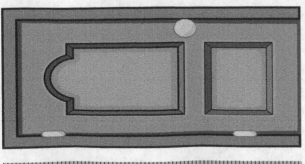

Line up one end of the measuring tape with the door. Make sure the edge of the door is at 0.

Look at the number that lines up with the other end of the door.

This door is 72 inches long. How many feet is that?
Hint: 12 inches = 1 foot!

think!
Since there are 12 inches in a foot, the door is 6 feet long.

Length Using Centimeters

A centimeter is a unit that can be used to measure length.

Measure this pen using a centimeter ruler.

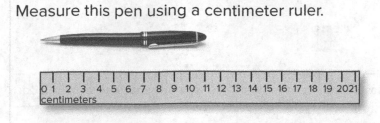

Line up one end of the pen at the 0 mark on the centimeter ruler.

Look at the number that lines up with the other end of the pen. This pen is 11 centimeters long.

PRACTICE: Now you try

Use inch units to measure the objects below.

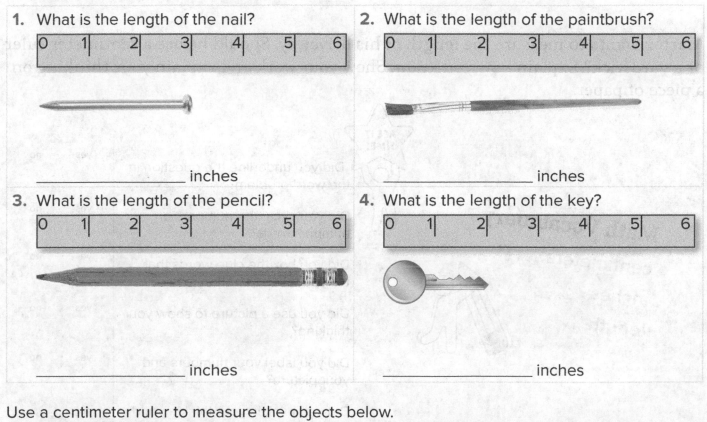

1. What is the length of the nail?

_____ inches

2. What is the length of the paintbrush?

_____ inches

3. What is the length of the pencil?

_____ inches

4. What is the length of the key?

_____ inches

Use a centimeter ruler to measure the objects below.

5. What is the length of the marker?

_____ centimeters

6. What is the length of the book?

Science ●●●

_____ centimeters

Use a centimeter ruler to measure the objects below.

7. What is the length of the phone?

8. What is the length of the water bottle?

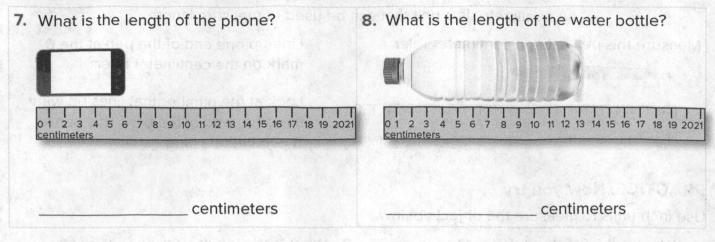

_____ centimeters

_____ centimeters

Martez wants to measure the length of his driveway. Should he use a centimeter ruler or a yardstick? Explain how you know. Show your work and explain your thinking on a piece of paper.

Math Vocabulary

centimeters

inches

length

ACE IT TIME!

	yes	no
Did you underline the question in the word problem?	○	○
Did you circle the numbers or number words?	○	○
Did you box the clue words that tell you what operation to use?	○	○
Did you use a picture to show your thinking?	○	○
Did you label your numbers and your picture?	○	○
Did you explain your thinking and use math vocabulary words in your explanation?	○	○

Math on the Move

Practice measuring objects around your home. Try measuring length with inch units. Also try to measure the length of objects using centimeter units.

Guess the Length

FOLLOWING THE OBJECTIVE
You will practice how to estimate (guess) length.

LEARN IT: You can estimate the length of an object in inches and in centimeters. Remember, an estimate is not an exact answer but rather a very good guess!

Example: About how many inches long is this screwdriver?

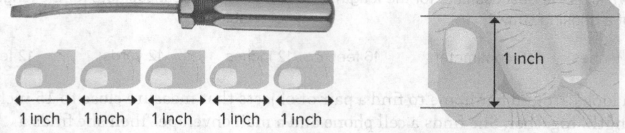

1 inch   1 inch   1 inch   1 inch   1 inch

1 inch

Hint: The distance from the tip of your thumb to the first knuckle is about an inch. You can use your thumb to estimate length. Look at the picture. Notice there are 5 thumbs lined up along the length of the screwdriver. You can see that this screwdriver is about 5 inches long.

Example: About how many centimeters long is this pencil?

Hint: The width of your pointer finger is about 1 centimeter. You can use the width of your finger to estimate length. Look at the picture. Notice there are 7 fingers lined up along the length of the pencil. You can see that this pencil is about 7 centimeters long.

PRACTICE: Now you try Estimate the length of each object in inches.

1. About how many inches long is the nail?

   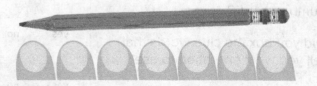

   _____ inches

2. About how many inches long is the paintbrush?

   _____ inches

Estimate the length of each object in centimeters.

3. About how many centimeters long is the leaf?

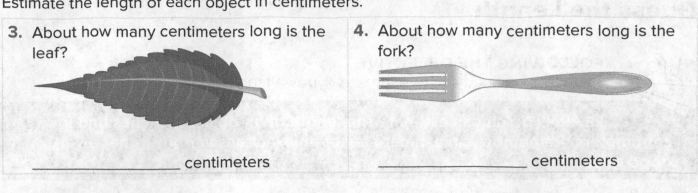

_____ centimeters

4. About how many centimeters long is the fork?

_____ centimeters

Circle the best estimate.

5. Which is the best estimate for the length of a drinking straw?

16 inches   16 centimeters   16 feet

6. Which is the best estimate for the length of a tablet?

12 inches   12 yards   12 feet

Rosa looks around her home to find a pair of objects that measure close to 15 inches in length, together. She finds a cell phone and a mail envelope. Then she finds a magazine and a notebook. Which pair of objects measures closer to 15 inches? Show your work and explain your thinking on a piece of paper.

ACE IT TIME!

	yes	no
Did you underline the question in the word problem?	yes	no
Did you circle the numbers or number words?	yes	no
Did you box the clue words that tell you what operation to use?	yes	no
Did you use a picture to show your thinking?	yes	no
Did you label your numbers and your picture?	yes	no
Did you explain your thinking and use math vocabulary words in your explanation?	yes	no

Math Vocabulary

estimate

inches

centimeters

Math on the Move

Practice estimating measurements around your home. Use your finger and thumb. Double-check your estimate using an inch or a centimeter ruler.

Measure the Same Length with Different Units

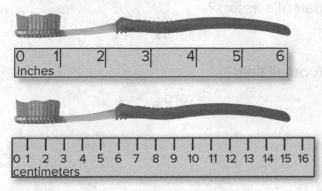

FOLLOWING THE OBJECTIVE
You will measure an object with two different units and compare.

LEARN IT: You can measure the same length with different types of units. In this lesson, we will use inches and centimeters to measure the length of an object. Use what you have learned about measuring length with rulers.

Example: Does it take more inches or centimeters to measure this toothbrush?

Use an <u>inch</u> ruler to measure this toothbrush.
This toothbrush measures 6 inches long.

Use a <u>centimeter</u> ruler to measure the same toothbrush
This toothbrush measures 15 centimeters long.

It takes more centimeters than inches to measure the length of this toothbrush. Can you explain why? *Hint:* The toothbrushes are the same size, but are inches and centimeters?

PRACTICE: Now you try Measure the objects below using the inch and centimeter rulers.

1.

_____ inches   _____ centimeters

So, it takes more _____ than _____ to measure the nail.

2.

_____ inches   _____ centimeters

So, it takes more _____ than _____ to measure the water bottle.

Unit 6: Measurement and Data Concepts

3.

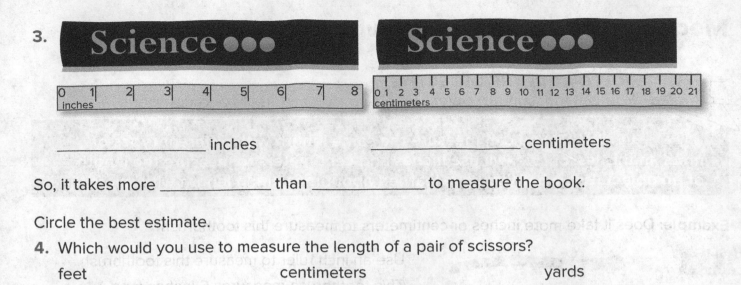

Science ●●●	Science ●●●

_____ inches   _____ centimeters

So, it takes more _____ than _____ to measure the book.

Circle the best estimate.

4. Which would you use to measure the length of a pair of scissors?

feet   centimeters   yards

5. Which would you use to measure the length of a football field?

inches   centimeters   yards

Zen measures the width of her desk in centimeters. She says that it is about 18 centimeters long. She also measures the width of her desk in inches. She says that it is about 24 inches long. Something is wrong with Zen's measurements. What is it? Show your work and explain your thinking on a piece of paper.

ACE IT TIME!

	yes	no
Did you underline the question in the word problem?	yes	no
Did you circle the numbers or number words?	yes	no
Did you box the clue words that tell you what operation to use?	yes	no
Did you use a picture to show your thinking?	yes	no
Did you label your numbers and your picture?	yes	no
Did you explain your thinking and use math vocabulary words in your explanation?	yes	no

Math on the Move

Use an inch ruler to measure objects around your home. Then use a centimeter ruler to measure those same objects. Have a conversation with an adult or a friend about what you notice.

Compare Lengths

FOLLOWING THE OBJECTIVE
You will measure to find how much longer one object is than another.

LEARN IT: You can compare the lengths of two objects. Measure the objects and then find how much longer one object is than the other.

Example: How many centimeters longer is the bottom pencil than the top pencil?

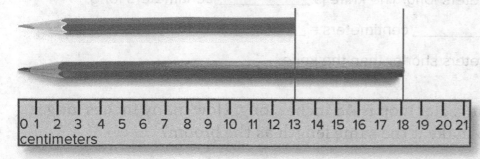

Step 1: Line up one end of each pencil with the 0 on the centimeter ruler.

Step 2: Measure. One pencil is 13 centimeters long. The other pencil is 18 centimeters long.

Step 3: Compare. Find the difference.

18 centimeters – 13 centimeters = 5 centimeters.

The bottom pencil is 5 centimeters longer than the top pencil.

PRACTICE: Now you try

1. How many inches shorter is the top piece of licorice than the bottom piece? Measure the length of each piece of licorice.

The top piece of licorice is _____ inches long.

The bottom piece of licorice is _____ inches long.

_____ inches – _____ inches = _____ inches.

The top piece of licorice is _____ inches shorter than the bottom piece.

2. How many centimeters shorter is the fork than the knife? Measure the lengths of the fork and the knife.

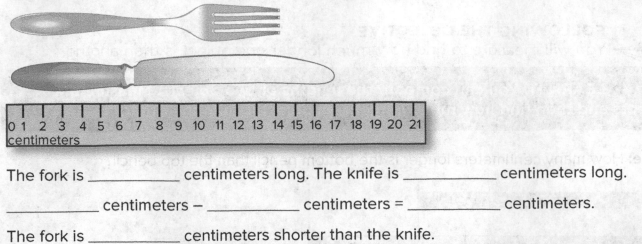

The fork is _____ centimeters long. The knife is _____ centimeters long.

_____ centimeters – _____ centimeters = _____ centimeters.

The fork is _____ centimeters shorter than the knife.

Katie's broom is 62 inches long. Her mop is 48 inches long. How many inches longer would the mop need to be to make it the same length as the broom?

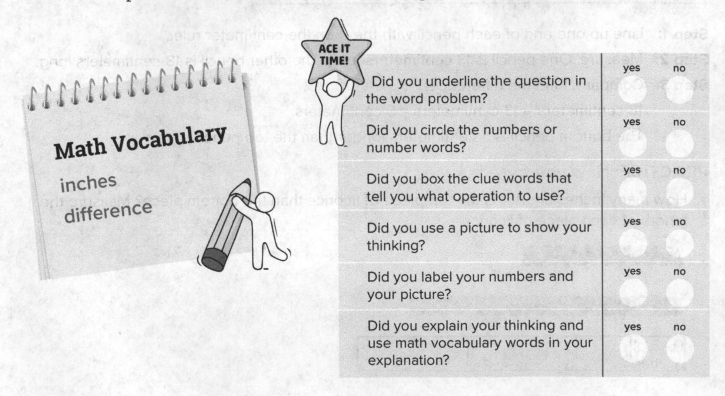

Math Vocabulary

inches

difference

ACE IT TIME!

	yes	no
Did you underline the question in the word problem?	◯	◯
Did you circle the numbers or number words?	◯	◯
Did you box the clue words that tell you what operation to use?	◯	◯
Did you use a picture to show your thinking?	◯	◯
Did you label your numbers and your picture?	◯	◯
Did you explain your thinking and use math vocabulary words in your explanation?	◯	◯

Math on the Move

Measure objects around your room and practice comparing the measurements. Write a number sentence to show the difference. You could also use inches to practice measuring two objects and comparing their lengths. Use centimeters to practice measuring two objects and comparing their lengths.

Solve Problems by Adding and Subtracting Lengths

FOLLOWING THE OBJECTIVE
You will add and subtract lengths to solve problems.

LEARN IT: You can add lengths to solve problems. You can subtract lengths to solve problems.

Example: Isabella has a piece of yarn that is 5 inches long and a piece of yarn that is 2 inches long. How many inches of yarn does she have in all?

5 inches   2 inches

You can use a picture to help you solve the problem. Add to find out how many inches of yarn Isabella has in all.

Write an equation: 5 + 2 = 7

You can also use a ruler to solve

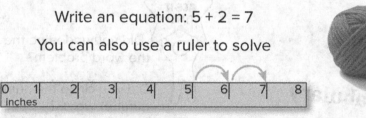

Start at 5. Remember, Isabella also has a piece of yarn that is 2 inches long. So we can hop forward 2 times on the ruler to show addition. When you hop forward 2 times, you land on 7 inches.

$$5 + 2 = 7$$

Example: Leanna has a pencil. When she started school in the morning, her pencil was 20 centimeters long. At lunch, she dropped her pencil and someone stepped on it. The pencil broke in two. The part that Leanna could write with was 9 centimeters shorter than it was in the morning. How long is her pencil now?

20 centimeters

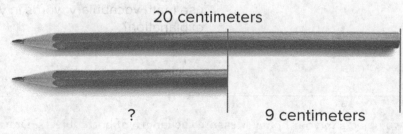

?   9 centimeters

You can use a picture to help you solve the problem. Subtract to find out how many centimeters long her pencil is now.

Write an equation: 20 − 9 = 11

PRACTICE: Now you try Write an equation to solve. You may also draw a picture or use a number line to help.

1. Lenny hopped 10 inches and then hopped another 12 inches. How far did Lenny hop?

2. Malcom swam 38 meters across the swimming pool. Lexi swam 13 meters less than Malcom. How many meters did Lexi swim?

3. Michael runs 34 yards across the football field. He is trying to get to the 50-yard line. How many more yards does he need to run to get to the 50-yard line?

The lamp in Mrs. Miller's living room is 55 inches tall. The lamp in her dining room is 12 inches taller. How tall is the lamp in Mrs. Miller's dining room? Show your work and explain your thinking on a piece of paper.

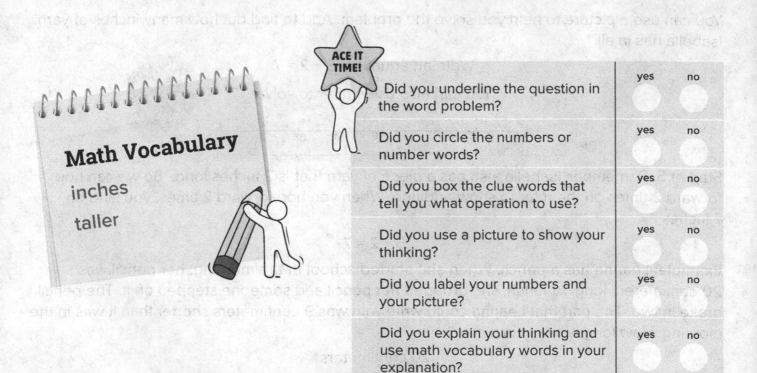

Math Vocabulary

inches

taller

ACE IT TIME!

	yes	no
Did you underline the question in the word problem?	○	○
Did you circle the numbers or number words?	○	○
Did you box the clue words that tell you what operation to use?	○	○
Did you use a picture to show your thinking?	○	○
Did you label your numbers and your picture?	○	○
Did you explain your thinking and use math vocabulary words in your explanation?	○	○

Math on the Move Find two objects in your home. Measure the length of both objects. Draw a number line to help you add the measurements of the two objects or to help you to find the difference in length between the two objects. Try this again with two other objects around your home.

Picture Graphs

FOLLOWING THE OBJECTIVE
You will read and make a picture graph.

LEARN IT: You can use data to make a picture graph. A *picture graph* uses pictures or symbols to show the value of data. A *key* tells what each picture stands for.

This picture graph shows the favorite after-school snacks for students in Mr. Wilson's class.

Favorite After-School Snacks for Students in Mr. Wilson's Class	
Grapes	♥ ♥ ♥ ♥ ♥
Apple Slices	♥ ♥ ♥ ♥
Yogurt	♥ ♥ ♥
Granola Bars	♥ ♥ ♥ ♥ ♥ ♥

Key: Each ♥ = 1 student

How many students picked yogurt as their favorite snack?

Look closely at the graph. Look at the row labeled "yogurt." Look at the key. The key shows that each heart represents one student. There are three hearts in the yogurt row, so three students picked yogurt as their favorite snack.

You can use data to make a picture graph.

Gloria made cookies for her birthday party. She made four chocolate chip cookies, five oatmeal raisin cookies, three peanut butter cookies, and three sugar cookies. First, complete the table by recording the **data** in the column labeled "number of cookies."

Cookies for the Party	
Type of Cookie	**Number of Cookies**
Chocolate Chip	4
Peanut Butter	
Oatmeal Raisin	
Sugar	

Use the table to make a picture graph.

The title of this picture graph is "Cookies for the Party." Label the rows with the different flavors of cookies.

Make a key. The key tells you that each symbol represents one cookie.

Draw cookie symbols to show the number of chocolate chip, peanut butter, oatmeal raisin, and sugar cookies.

Cookies for the Party	

Key: Each ___ = 1 cookie

PRACTICE: Now you try Use the data to complete the table, make a graph, and then answer questions 3–6.

1. Derrick read many of his favorite books last summer. In May he read four books, in June he read six books, in July he read eight books, and in August he read five books.

Books Derrick Read	
Month	**Number of Books**

2. Use the table to make a picture graph. Fill in the title of your graph. Label the rows. Draw your symbols to show the number of books. Draw the key below.

Title:	

3. In which month did Derrick read the least amount of books?

_____

4. In which month did Derrick read the most books?

_____

5. How many more books did Derrick read in July than he did in May?

_____

6. How many books did Derrick read in July and August all together?

_____

Use the graph "Cookies for the Party." How many more oatmeal raisin and chocolate chip cookies combined are there than sugar cookies and peanut butter cookies combined? Show your work and explain your thinking on a piece of paper.

Math Vocabulary

data

picture graph

key

difference

ACE IT TIME!

	yes	no
Did you underline the question in the word problem?	yes	no
Did you circle the numbers or number words?	yes	no
Did you box the clue words that tell you what operation to use?	yes	no
Did you use a picture to show your thinking?	yes	no
Did you label your numbers and your picture?	yes	no
Did you explain your thinking and use math vocabulary words in your explanation?	yes	no

Math on the Move

Think of a survey question to ask family members and/or friends. For example, "What is your favorite ice cream flavor?" or "What is or your favorite sport?" Collect their responses and create a table and a picture graph to represent their responses.

Bar Graphs

FOLLOWING THE OBJECTIVE
You will read and make a bar graph.

LEARN IT: A *bar graph* is another way we can look at data, or information. Instead of using pictures, it uses bars to show the value of the data. Let's use the information we read about in our last lesson on picture graphs and put that same information into a bar graph.

This bar graph shows the favorite after-school snacks for students in Mr. Wilson's class.

How many students picked granola bars as their favorite snack?

Notice the scale on the left side of the graph. It goes from 0 to 7. Look closely at the graph. Look at the column, or bar, labeled "Granola Bars." Look at the top of the bar. It lines up with the number 6 on the scale. So six students picked granola bars as their favorite snack.

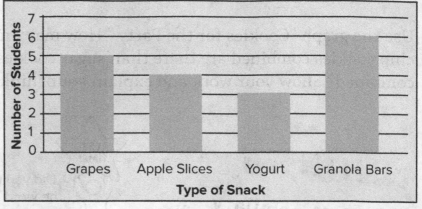

PRACTICE: Now you try

Step 1: Use the data to complete the bar graph.

Remember the table you made last lesson about Derrick's summer reading? Use the information from that table to create a bar graph.

Books Derrick Read	
Month	**Number of Books**
May	4
June	6
July	8
August	5

Books Derrick Read

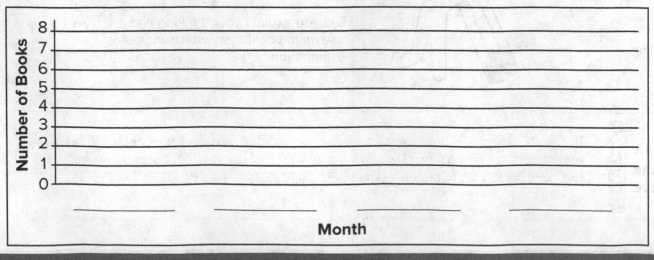

Step 2. Use your bar graph on page 90 to answer the questions below.

1. In which month did Derrick read the most books?

_____

2. In which month did Derrick read the least amount of books?

_____

3. How many more books did Derrick read in July than he did in August?

_____

4. How many books did Derrick read in July and May all together?

_____

Use the bar graph "Books Derrick Read" on page 90. Did Derrick read more books in the first part of the summer (May and June combined) or did he read more books in the second part of the summer (July and August combined)? How many more? Show your work and explain your thinking on a piece of paper.

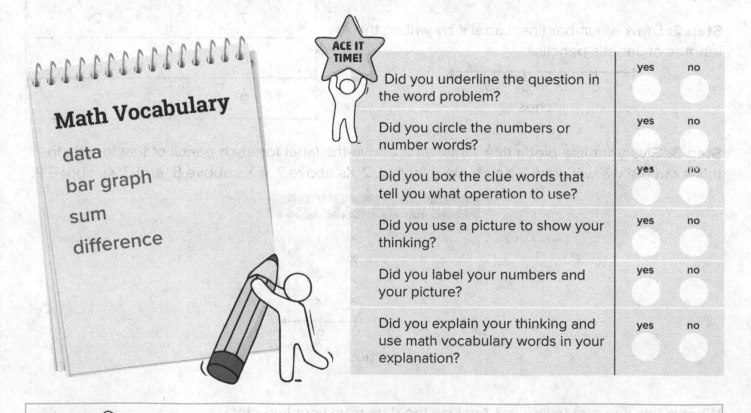

Math Vocabulary

data

bar graph

sum

difference

ACE IT TIME!

	yes	no
Did you underline the question in the word problem?	○	○
Did you circle the numbers or number words?	○	○
Did you box the clue words that tell you what operation to use?	○	○
Did you use a picture to show your thinking?	○	○
Did you label your numbers and your picture?	○	○
Did you explain your thinking and use math vocabulary words in your explanation?	○	○

Math on the Move
Practice creating bar graphs! Count how many pencils, pens, and scissors you can find in your home. Create a table and a bar graph with your data. Can you think of other objects from around your home or classrooom you could count and graph?

Line Plots

FOLLOWING THE OBJECTIVE
You will understand and create line plots to display a data set of measurements. You will make observations about the data.

LEARN IT: A *line plot* is another way to visually represent data. A line plot records the number of times (frequency) data occurs. It shows the frequency of data along a number line.

Example: Janet has 11 pencils in her pencil case. The pencils are different lengths, as shown in the table here. You can make a line plot to show the lengths of Janet's pencils.

Step 1: Look at the data in the table.

Step 2: Draw a number line. Label it by writing the lengths of Janet's pencils.

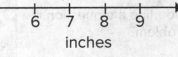

inches

Length of Janet's Pencils	
Length (in inches)	**Number of Pencils**
6	3
7	2
8	4
9	2

Step 3: Give your line plot a title. Draw an X above the label for each pencil of that length. In this example we will draw 3 Xs above 6 inches, 2 Xs above 7, 4 Xs above 8, and 2 Xs above 9.

Length of Janet's Pencils

```
                    X
      X             X
      X    X    X   X
      X    X    X   X
   ←——6————7————8————9——→
           inches
```

What is this line plot telling us? Analyze the data from your line plot.
There are 11 Xs, so there are 11 pencils all together.
(Each X represents 1 pencil.)

Use the line plot of Janet's pencils to answer these questions:

1. How many pencils are 8 inches or greater?
   4 (height of 8 inches) + 2 (height of 9 inches) = 6 pencils

2. How many pencils are 6 inches long? 3 pencils

3. What is the difference between the lengths of the longest pencil and the shortest pencil?
   9 inches − 6 inches = 3 inches

PRACTICE: Now you try

Use the data to complete the table. Complete the line plot. Answer the questions for the line plot.

The students in Mrs. Ortega's class had to walk one mile in gym class. The data shows the length of time in minutes it took her class to walk one mile. (14, 14, 14, 15, 15, 15, 15, 16, 16, 17, 17)

Step 1: Record the number of students in the table.

Length of Time Students Walked in Mrs. Ortega's Class	
Time (in Minutes)	Number of Students
14	
15	
16	
17	

Step 2: Label the number line below with the times it took Mrs. Ortega's class to walk one mile.

Step 3: Display the data. Draw an X above the label for the time it took each student. Write the title of the line plot above it.

Title:

⟵───────────────────────⟶

Time (in Minutes)

Unit 6: Measurement and Data Concepts

After creating the line plot, use the information to answer the questions:

1. How many total students are in Mrs. Ortega's class? _____

2. How many more students walked for 15 minutes than for 17 minutes? _____

3. How many students walked for 15 minutes or more? _____

4. How many students walked for 16 minutes or more? _____

Use the line plot "Length of Time Students Walked in Mrs. Ortega's Class" on page 93 to answer the following question: Which is greater, the number of students that walked 15 minutes or less or the number of students that walked 16 minutes or more? Show your work and explain your thinking on a piece of paper.

Math Vocabulary

sum

total

difference

subtract

compare

line plot

data

ACE IT TIME!

	yes	no
Did you underline the question in the word problem?	○	○
Did you circle the numbers or number words?	○	○
Did you box the clue words that tell you what operation to use?	○	○
Did you use a picture to show your thinking?	○	○
Did you label your numbers and your picture?	○	○
Did you explain your thinking and use math vocabulary words in your explanation?	○	○

Math on the Move

Create a line plot. Count the number of letters in each family member's first name and plot them. Make up questions about your line plot for your family members to answer.

REVIEW

Congratulations! You've finished the lessons for Units 5–6. This means you've learned how to solve word problems with different kinds of money, tell and write time, estimate and measure the length of objects using various units, show measurements by making a line plot, and now you can even use addition and subtraction within 100 to solve word problems involving lengths.

Now it's time to prove your skills with measurement and data concepts. Solve the problems below. Use all the skills you have learned.

Activity Section 1

1. Write the time on the digital clock. Circle A.M. or P.M.

   Eat an after-school snack

   A.M.   P.M.

2. Use the time on the digital clock to draw in the short and long hands. Circle A.M. or P.M.

   Getting ready for bed

   A.M.   P.M.

   8 : 15

Activity Section 2 Solve. You may draw coins to help you solve.

1. Wendy has 4 nickels, 6 pennies, and 1 quarter in her piggy bank. How much money does she have?

2. Donna found 4 quarters, 2 dimes, 4 nickels, and 4 pennies in her dad's car. How much money did she find?

Use only coins to show the amount in two ways. Draw and label the coins that you choose.

3.

$1.34

Activity Section 3 Use the given tool to measure the objects below.

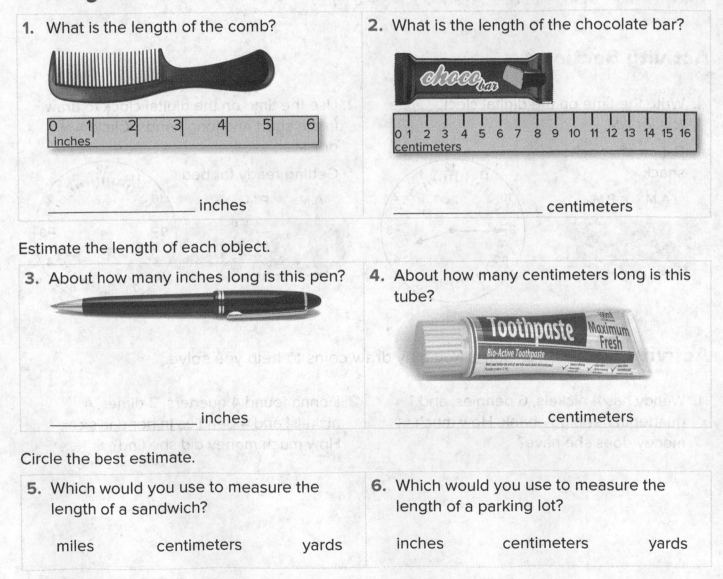

1. What is the length of the comb?

_____ inches

2. What is the length of the chocolate bar?

_____ centimeters

Estimate the length of each object.

3. About how many inches long is this pen?

_____ inches

4. About how many centimeters long is this tube?

_____ centimeters

Circle the best estimate.

5. Which would you use to measure the length of a sandwich?

miles   centimeters   yards

6. Which would you use to measure the length of a parking lot?

inches   centimeters   yards

Activity Section 4 Write an equation to solve. You may also draw a picture or use a number line to help.

1. Nevil hopped 21 inches, and his brother Jeremy hopped 27 inches. How far did they hop all together?

Activity Section 5 Answer the questions for the picture graph "Favorite Vegetable."

1. How many students selected corn as their favorite vegetable? _____

2. How many more students chose carrots than chose potatoes? _____

3. How many students chose peas and corn all together? _____

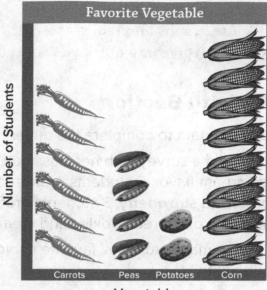

Favorite Vegetable

Number of Students

Carrots   Peas   Potatoes   Corn

Vegetable

Key: Each symbol = one student

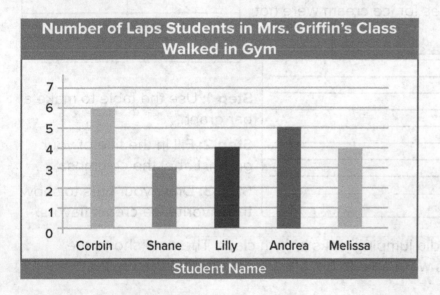

Number of Laps Students in Mrs. Griffin's Class Walked in Gym

Corbin   Shane   Lilly   Andrea   Melissa

Student Name

Answer the questions for the bar graph "Number of Laps Walked in Gym."

4. How many more laps did Corbin walk than Shane? _____

5. Which two students walked the same number of laps? _____ and _____

6. How many laps did Andrea and Lilly walk all together? _____

Answer the questions for the line plot.

7. How many students had 5 pencils in their desks? _____

8. How many more students had 5 pencils than had 8 pencils in their desks? _____

9. How many students had 8 or more pencils in their desks? _____

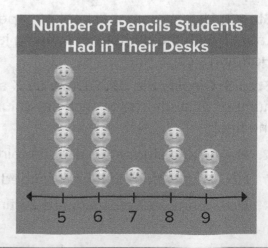

Number of Pencils Students Had in Their Desks

5   6   7   8   9

UNDERSTAND

Understand the meaning of what you have learned and apply your knowledge.

We use graphs and tables as a quick way to look at data. Picture graphs, bar graphs, and line plots give us a quick view of the "big picture!"

Activity Section

Use the data to complete the table.

Lydia did a survey with her classmates to find out their favorite ice cream flavor. 5 students preferred chocolate, 7 students preferred strawberry, 3 students preferred vanilla, and 4 students preferred cookies and cream.

1. How many students' favorite choices for ice cream were not chocolate?

Favorite Ice Cream Flavor	
Ice Cream Flavor	Number of Students

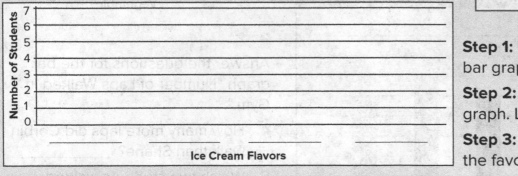

Step 1: Use the table to make a bar graph.

Step 2: Fill in the title of your graph. Label the columns.

Step 3: Draw your bars to show the favorite ice cream flavors.

The students in Mrs. Hoffman's class did jumping jacks in gym class. The data shows the number of jumping jacks the students were able to do. (24, 24, 24, 25, 25, 25, 25, 25, 26, 26, 27, 27).

Step 1: Complete the table and answer the questions.

Step 2: Use a separate sheet of paper to create a line plot. Label it with the number of jumping jacks the students in Mrs. Hoffman's class did.

Step 3: Display the data. Draw an X above the label for the number of students for each set of jumping jacks. Don't forget the title!

2. How many students did 24 jumping jacks?

3. How many more students jumped 25 times than 27 times?

4. How many students jumped 26 times or more?

Number of Jumping Jacks Done in Mrs. Hoffman's Class	
Number of Jumping Jacks	Number of Students
24	
25	
26	
27	

DISCOVER

Counting money is a real-world skill that you will use for the rest of your life! Once you are comfortable with skip counting, counting money becomes easier.

Activity Section

Che has some quarters, dimes, nickels, and pennies. The money he has is worth 73¢. He spends 28¢ for a colored pen at the book fair.

1. What is the least number of coins that Che could have left? _____

   Draw the coins that he has left.

   Explain how you know this is the least number of coins he could have left.

2. What is the most amount of coins that Che could have left? _____

   Draw the coins that he has left.

   Explain how you know this is the most amount of coins he could have left.

Geometry Concepts

Polygons and Shapes

FOLLOWING THE OBJECTIVE
You will identify *polygons* and the vertices and angles that they have.

LEARN IT: The world is made up of *polygons*, or shapes. Polygons are shapes that have at least three sides and have all the sides touching each other. The points where the sides meet are called vertices. The vertices form *angles*.

Think of a *polygon* as a fence around a yard. The fence defines the shape of the yard. Think of the vertices as the corners in the yard. The chart below describes some popular polygons.

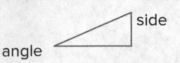

Polygons	Name	Number of Sides	Number of Vertices and Angles
△	Triangle	3	3
□	Quadrilateral (Square)	4	4
▭	Quadrilateral (Rectangle)	4	4
⬠	Pentagon	5	5
⬡	Hexagon	6	6

These shapes are *not polygons*.
The have curved sides.

These shapes are *not polygons*.
All their sides do not touch each other.

PRACTICE: Now you try Name these polygons.
Tell how many sides and how many angles they each have.

1.

Name _____

Sides _____

Angles _____

2.

Name _____

Sides _____

Angles _____

3.

Name _____

Sides _____

Angles _____

4.

Name _____

Sides _____

Angles _____

Explain how these two polygons are the same and how they are different. Show your work and explain your thinking on a piece of paper.

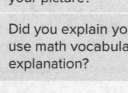

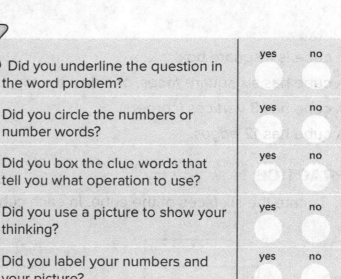

ACE IT TIME!

	yes	no
Did you underline the question in the word problem?		
Did you circle the numbers or number words?		
Did you box the clue words that tell you what operation to use?		
Did you use a picture to show your thinking?		
Did you label your numbers and your picture?		
Did you explain your thinking and use math vocabulary words in your explanation?		

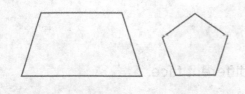

Math Vocabulary

quadrilateral

angle

pentagon

vertices

sides

Math on the Move You can find different shapes all around your home or classroom! Find three examples of each. For example, a rug might be a quadrilateral or a hexagon. What shape is your refrigerator?

The Cube

FOLLOWING THE OBJECTIVE
You will identify a cube by knowing all the parts that make up a cube.

LEARN IT: A *cube* is a three-dimensional shape. It is a solid shape. Let's see what makes a cube a cube!

This is a square.	Think of 2 squares standing up on their edges.	Think of connecting the corners of the squares. Now you have made a cube!

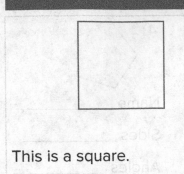

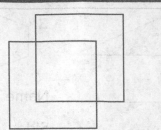

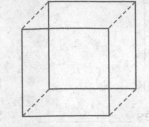

A cube is a square box.

A cube has six square *faces*, or sides: the top, the bottom, and four sides.

A cube has 8 *vertices* (corners).

A cube has 12 *edges*.

PRACTICE: Now you try

1. Locate the six faces of the cube. In each cube, color a different face.

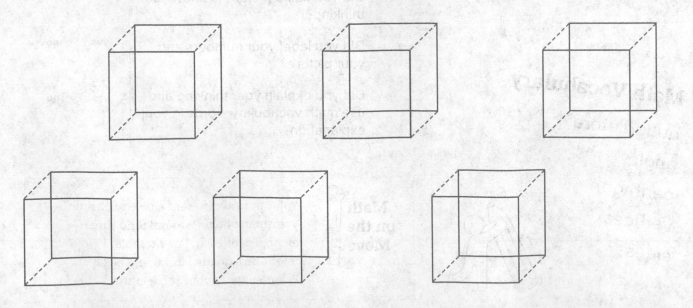

2. Place a dot on each vertex or corner of the cube. Prove there are eight. Count each edge by putting a number on it.

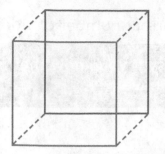

Miss Franklin asks her students to compare these two shapes. Miguel says they are both cubes. Jaxon says only one is a cube. Who do you agree with and why? Show your work and explain your thinking on a piece of paper.

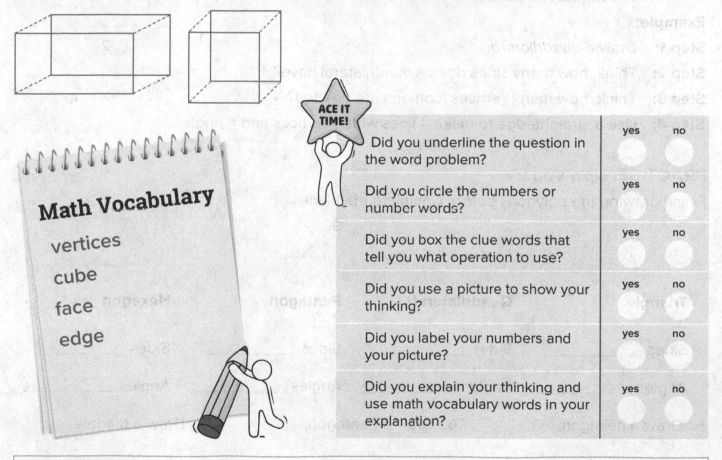

Math Vocabulary

vertices

cube

face

edge

ACE IT TIME!

	yes	no
Did you underline the question in the word problem?	○	○
Did you circle the numbers or number words?	○	○
Did you box the clue words that tell you what operation to use?	○	○
Did you use a picture to show your thinking?	○	○
Did you label your numbers and your picture?	○	○
Did you explain your thinking and use math vocabulary words in your explanation?	○	○

Math on the Move

Make a list of cubes that you can find in the real world. How many cubes can you find in your home?

Let's Draw Polygons

FOLLOWING THE OBJECTIVE
You will draw *polygons* with the correct number of sides and angles.

LEARN IT: Drawing *polygons* is easy if you know their names and the number of sides, vertices, and angles they have.

Step 1: Know which polygon you have to draw.

Step 2: Think of how many sides the polygon has.

Step 3: Think of how many *vertices* (corners) the polygon has.

Step 4: Use a straightedge or a ruler to draw your lines. Remember that all the lines have to connect to each other.

Example:

Step 1: Draw a *quadrilateral*.

Step 2: Think, how many sides does a quadrilateral have? 4

Step 3: Think, how many vertices (corners) are needed? 4

Step 4: Use a straightedge to make 4 lines with 4 vertices and 4 angles.

PRACTICE: Now you try

Finish drawing the polygons below. Complete the blanks.

1.

Triangle

Sides _____

Angles _____

2.

Quadrilateral

Sides _____

Angles _____

3.

Pentagon

Sides _____

Angles _____

4.

Hexagon

Sides _____

Angles _____

5. Draw a hexagon.

6. Draw a pentagon.

7. Draw a triangle.

8. Draw three different quadrilaterals.

Design a class flag. The flag must have two quadrilaterals, a hexagon, a pentagon, and two triangles. Label each part. Show your work and explain your thinking on a piece of paper.

Math Vocabulary

quadrilateral

angle

triangle

hexagon

pentagon

vertices

ACE IT TIME!

	yes	no
Did you underline the question in the word problem?		
Did you circle the numbers or number words?		
Did you box the clue words that tell you what operation to use?		
Did you use a picture to show your thinking?		
Did you label your numbers and your picture?		
Did you explain your thinking and use math vocabulary words in your explanation?		

Math on the Move

Most countries have their own national flags. With the permission of an adult, go online and look up some flags from other countries. Look for different kinds of polygons in each flag.

The Great Rectangle Divide

FOLLOWING THE OBJECTIVE
You will be able to break apart a rectangle into rows and columns of squares that are the same size.

LEARN IT: How do you find out how many same-sized squares will cover a rectangle? You can imagine you are breaking apart the rectangle into smaller squares and count to find out how many squares can cover the rectangle. Let's look at this rectangle. How many of these gray squares would fit inside?

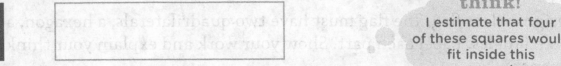

think!
I estimate that four of these squares would fit inside this rectangle.

Step 1: Compare the size of the square and the size of the rectangle. Guess how many squares will fit. This is called an estimate.

Step 2: Imagine you could trace the square and cut it out. Imagine placing it on top of the rectangle on one end. Draw a line to show where the square would fit.

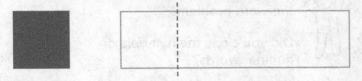

Step 3: Imagine adding more squares on top of the rectangle until the whole rectangle has been filled. Draw lines to show where the squares would fit. (You can see that four squares would fit!)

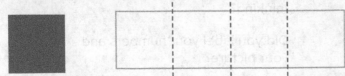

Let's look at these two rectangles below. They both have eight squares that fit inside of them, but they have a different number of *rows* and *columns*. Remember, rows go side to side and columns go up and down!

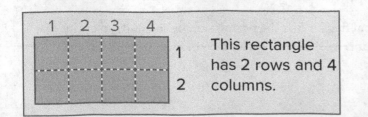

This rectangle has 2 rows and 4 columns.

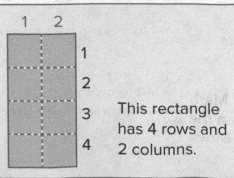

This rectangle has 4 rows and 2 columns.

PRACTICE: Now you try

1. Draw how many squares will fit inside these rectangles. Tell how many rows and how many columns.

_____ squares _____ rows _____ columns

2.

_____ squares _____ rows _____ columns

McKenna draws a rectangle with two rows and three columns. The squares inside the rectangle are all the same size. What could her rectangle look like? Show your work and explain your thinking on a piece of paper.

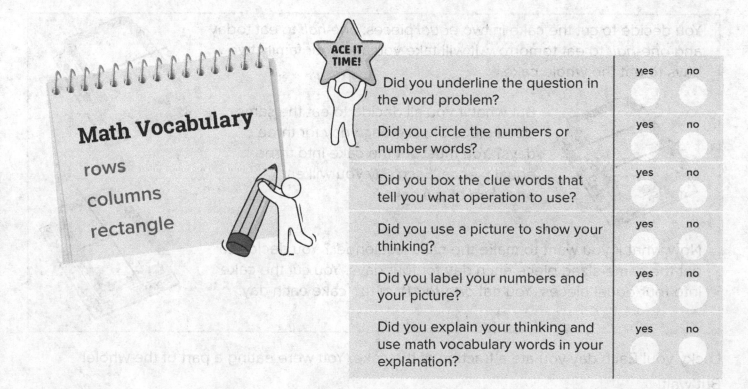

Math Vocabulary

rows

columns

rectangle

ACE IT TIME!

	yes	no
Did you underline the question in the word problem?	○	○
Did you circle the numbers or number words?	○	○
Did you box the clue words that tell you what operation to use?	○	○
Did you use a picture to show your thinking?	○	○
Did you label your numbers and your picture?	○	○
Did you explain your thinking and use math vocabulary words in your explanation?	○	○

Math on the Move

Cut out a square. Find a rectangle shape in your home, like a picture or a tabletop. See how many of the squares that you cut out will cover the space. How many rows and columns can you cover?

Fraction Concepts

What Is a Fraction?

FOLLOWING THE OBJECTIVE
You will identify that a fraction is an equal part of a whole.

LEARN IT: A fraction is a number that expresses parts of a group or a whole. Fractions help when you need to measure things. Measurements will not always equal a whole number.

Example: Your mother buys a surprise cake for your family. You and your family decide you will not eat the *whole* cake in one day.

You decide to cut the cake in two *equal* pieces: *one-half* to eat today and *one-half* to eat tomorrow. It will take you and your family two days to eat the whole cake.

But what if you all decide to eat the same-sized piece of cake each day for three days? You must cut the cake into three equal pieces. Each day you will eat *one-third* of the cake.

Now what if you want to make the cake last longer? You decide to eat the same-sized piece each day for four days. You cut the cake into four equal pieces. You eat *one-fourth* of the cake each day.

Lucky you! Each day you ate a fraction of the cake. You were eating a part of the whole!

But wait!

Fractions must be *equal shares*. Do the pictures below show fractions? No! They are NOT fractions because the parts are not equal.

 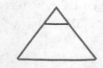

PRACTICE: Now you try

1. Circle the shapes that show a fraction. Place an X on the shapes that do not show fractions.

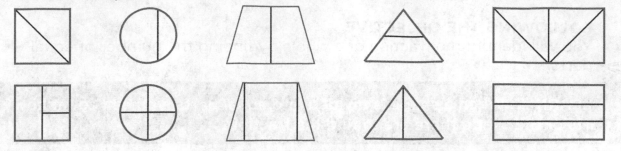

2. Color the shapes above that show halves, thirds, and fourths.

Zia cut a foot-long sandwich into thirds to share with her friends. How many equal pieces of sandwich did she make? Draw a picture to prove your answer. Show your work and explain your thinking on a piece of paper.

Math Vocabulary

whole

equal share

one-third

fraction

ACE IT TIME!

	yes	no
Did you underline the question in the word problem?	○	○
Did you circle the numbers or number words?	○	○
Did you box the clue words that tell you what operation to use?	○	○
Did you use a picture to show your thinking?	○	○
Did you label your numbers and your picture?	○	○
Did you explain your thinking and use math vocabulary words in your explanation?	○	○

Math on the Move Look for recipes in a cookbook. Can you find recipes that use fractions? What do you think those fractions mean? Why are fractions so important in cooking?

Name the Fraction

FOLLOWING THE OBJECTIVE
You will identify the fraction of a whole by finding the number of equal parts.

LEARN IT: When a shape is divided into *equal* parts, you can count how many parts of the shape there are to find the fraction.

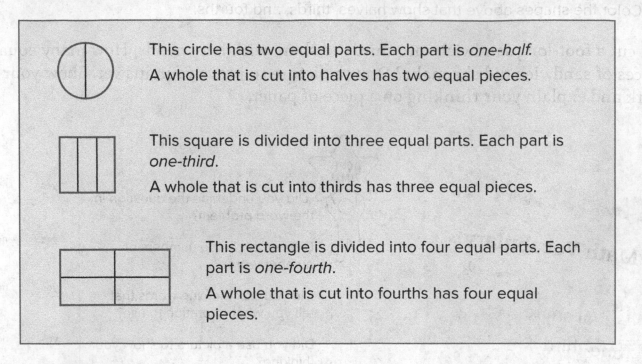

This circle has two equal parts. Each part is *one-half*.
A whole that is cut into halves has two equal pieces.

This square is divided into three equal parts. Each part is *one-third*.
A whole that is cut into thirds has three equal pieces.

This rectangle is divided into four equal parts. Each part is *one-fourth*.
A whole that is cut into fourths has four equal pieces.

PRACTICE: Now you try

1. Divide these shapes into halves.

2. Divide these shapes into fourths.

3. Divide these shapes into thirds.

Name the fractions. Tell how many equal parts.

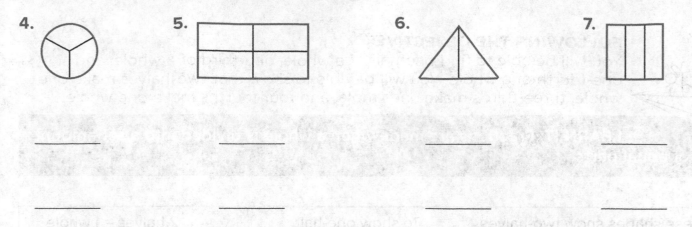

4. _____

    _____

5. _____

    _____

6. _____

    _____

7. _____

    _____

Neal cut a watermelon slice so he could share it with his three friends. He was so careful to make sure he made three equal shares. After he passed out the pieces, he found out there wasn't enough for him! What did he do wrong? Show your work and explain your thinking on a piece of paper.

Math Vocabulary

whole

equal share

thirds

fourths

fraction

ACE IT TIME!

	yes	no
Did you underline the question in the word problem?	○	○
Did you circle the numbers or number words?	○	○
Did you box the clue words that tell you what operation to use?	○	○
Did you use a picture to show your thinking?	○	○
Did you label your numbers and your picture?	○	○
Did you explain your thinking and use math vocabulary words in your explanation?	○	○

Math on the Move

Look around your home to find shapes that can be divided into equal parts. Make a drawing of them and show the fraction. Label the fraction.

Fraction Action with Rectangles and Circles

FOLLOWING THE OBJECTIVE
You will be able to find one-half of a whole, one-third of a whole, and one-fourth of a whole. You will be able to show that two-halves make one whole, three-thirds make one whole, and four-fourths make one whole.

LEARN IT: Fractions are made when a whole is divided into a set amount of equal parts.

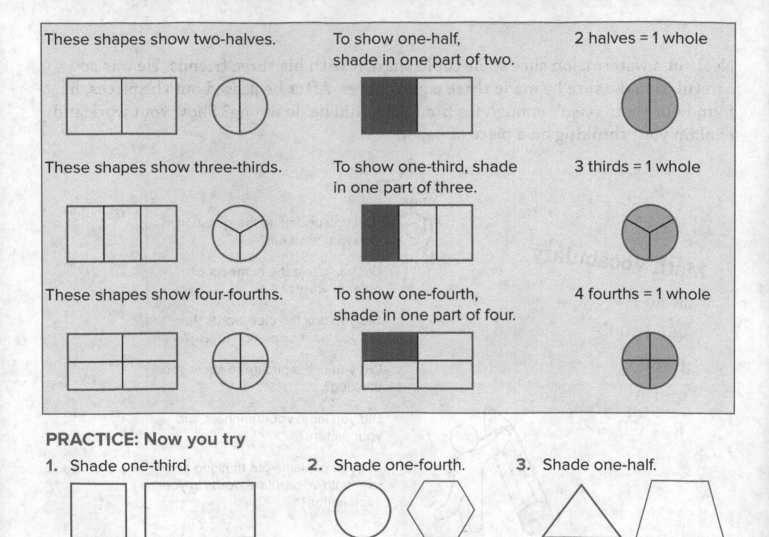

These shapes show two-halves.

To show one-half, shade in one part of two.

2 halves = 1 whole

These shapes show three-thirds.

To show one-third, shade in one part of three.

3 thirds = 1 whole

These shapes show four-fourths.

To show one-fourth, shade in one part of four.

4 fourths = 1 whole

PRACTICE: Now you try

1. Shade one-third.

2. Shade one-fourth.

3. Shade one-half.

4. It takes _____ thirds to make one whole. Draw a picture to show this.

5. It takes _____ halves to make one whole. Draw a picture to show this.

6. It takes _____ fourths to make one whole. Draw a picture to show this.

Clay had five friends over for his birthday party. His mom cut the cake into six equal pieces, one for each friend and one for Clay. Did they eat the whole cake? Show your work and explain your thinking on a piece of paper.

Math Vocabulary

whole

equal share

sixths

fraction

ACE IT TIME!

	yes	no
Did you underline the question in the word problem?	○	○
Did you circle the numbers or number words?	○	○
Did you box the clue words that tell you what operation to use?	○	○
Did you use a picture to show your thinking?	○	○
Did you label your numbers and your picture?	○	○
Did you explain your thinking and use math vocabulary words in your explanation?	○	○

Math on the Move Find out how pizzas are cut into fractions. Draw a picture to show that you ate one piece. How many pieces would you eat if you ate the whole thing?

Are Equal Shares Equal Fractions?

FOLLOWING THE OBJECTIVE
You will recognize that when you have shapes that are the same size and they are divided into the same number of equal shares, the shares do not have to have the same shape.

LEARN IT: Equal shares do not have to be the same shape. Shares must have the same fractional value, but they do not have to look the same.

These same-size rectangles are all divided into halves.

Each shaded half has a different shape, but it is still equal to one-half!

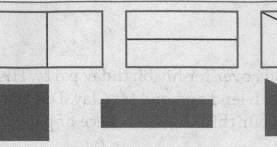
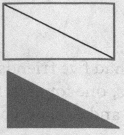

These same-size rectangles are divided into fourths.

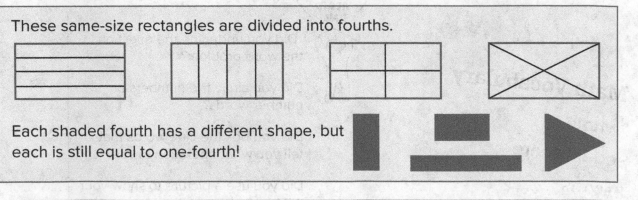

Each shaded fourth has a different shape, but each is still equal to one-fourth!

Look at these hexagon halves.

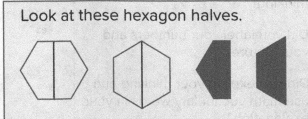

The shapes are the same size, but the halves are different shapes. Can you explain why? *Hint:* It has something to with where the hexagon is divided!

PRACTICE: Now you try

1. Find two different ways to divide this same-size shape into halves so that the halves will be different shapes.

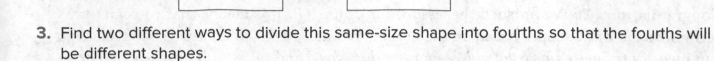

2. Find two different ways to divide this same-size shape into thirds so that the thirds will be different shapes.

3. Find two different ways to divide this same-size shape into fourths so that the fourths will be different shapes.

Mom made two peanut butter sandwiches that were the same-size squares. She wanted to cut each sandwich into fourths but wanted to make different shapes in each sandwich. How could she do this? Draw a picture to help. Show your work and explain your thinking on a piece of paper.

Math Vocabulary

whole

equal share

one-fourth

fraction

ACE IT TIME!

	yes	no
Did you underline the question in the word problem?		
Did you circle the numbers or number words?		
Did you box the clue words that tell you what operation to use?		
Did you use a picture to show your thinking?		
Did you label your numbers and your picture?		
Did you explain your thinking and use math vocabulary words in your explanation?		

Math on the Move

What shapes can you find that can only be divided into two equal shares in one way? What shapes can you find that can only be divided into three equal shares one way?

REVIEW

Stop and think about what you have learned.

Congratulations! You've finished the lessons for Units 7–8. This means that you can recognize and draw shapes based on the number of *lines*, *angles*, and *faces* a shape has. You can identify *triangles*, *quadrilaterals*, *pentagons*, *hexagons*, and *cubes*. You have learned how to divide shapes into fair shares and describe them using the words *half*, *third*, *fourth*, and *whole*. You can take a rectangle and divide it into equal-size squares and say how many squares there are in that rectangle.

Now it is time to prove what you know about shapes and fractions. Solve the problems below. Use all that you have learned in Units 7 and 8!

Activity Section 1 Fill in the blanks and then draw the shape.

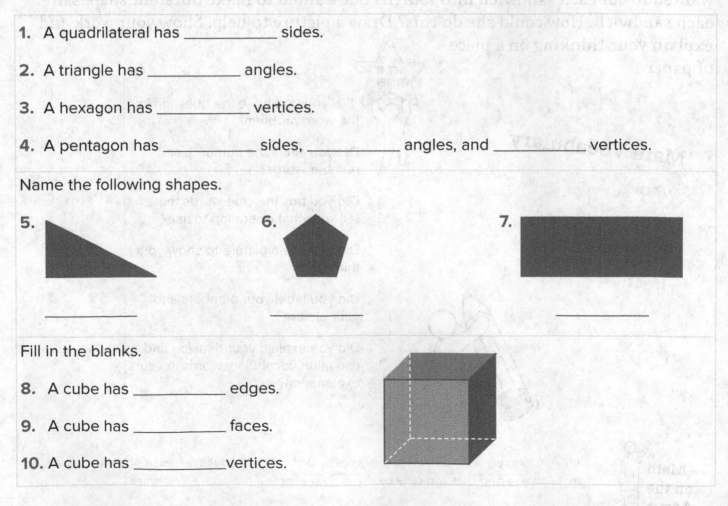

1. A quadrilateral has _____ sides.

2. A triangle has _____ angles.

3. A hexagon has _____ vertices.

4. A pentagon has _____ sides, _____ angles, and _____ vertices.

Name the following shapes.

5. _____

6. _____

7. _____

Fill in the blanks.

8. A cube has _____ edges.

9. A cube has _____ faces.

10. A cube has _____ vertices.

Activity Section 2 Answer the questions.

1. How many columns appear? _____

2. How many rows do you see? _____

3. How many squares fit in the rectangle? _____

4. Use this shape to draw a rectangle: ▪
   Draw four rows. Draw two columns.

   How many squares are there? _____

5. Draw a shape that has three rows and four columns. How many squares does the shape have?

Activity Section 3

1. Color the shapes cut in halves yellow.

   Color the shapes cut in thirds red.

   Color the shapes cut in fourths green.

Draw a shape showing:

2. Thirds

3. Fourths

4. Halves

Show the following by drawing a shape and shading the fraction.

5. One-third

6. One-fourth

7. One-half

8. Draw two rectangles the same size and show two different shapes that one-fourth can have.

UNDERSTAND

Understand the meaning of what you have learned and apply your knowledge.

Activity Section

Complete this table.

Number of Items	Fractional Part	Picture	How Much Is the Fraction?
12 eggs	one-third **think!** How many groups do I need?	X X X X X X X X X X X X	one-third of 12 = 4
10 marbles	one-half		
15 raisins	one-third		
12 apples	one-fourth		

DISCOVER

Let's see if you can take what you have learned to solve a problem about shapes and fractions.

Activity Section

Ashton's father orders a pizza for his birthday party. He can have the pizza cut into either six equal pieces or eight equal pieces. His friend Martin says he thinks the pizza should be divided into eight equal pieces because that will give everyone the largest-sized piece. Is he right? Explain your thinking with words and pictures.

Answer Key

Unit 2: Number Concepts

Odd or Even?

Page 9 Practice: Now you try

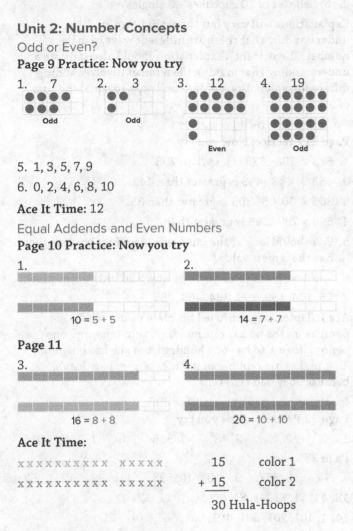

1. 7 — Odd
2. 3 — Odd
3. 12 — Even
4. 19 — Odd

5. 1, 3, 5, 7, 9
6. 0, 2, 4, 6, 8, 10

Ace It Time: 12

Equal Addends and Even Numbers

Page 10 Practice: Now you try

1. 10 = 5 + 5
2. 14 = 7 + 7

Page 11

3. 16 = 8 + 8
4. 20 = 10 + 10

Ace It Time:

x x x x x x x x x x   x x x x x   15   color 1
x x x x x x x x x x   x x x x x   + 15   color 2
                                   30 Hula-Hoops

There are 14 girls and 15 red Hula-Hoops. The girls can all have the same color. (There will be one red Hula-Hoop left over.)

There are 16 boys and 15 green Hula-Hoops. One boy will not get the green color.

Skip Counting by 5s, 10s, and 100s

Page 12

1. Hundreds Chart

    5s: (circle) 5, 10, 15, 20, 25, 30, 35, 40, 45, 50, 55, 60, 65, 70, 75, 80, 85, 90, 95, 100

    10s: (X) 10, 20, 30, 40, 50, 60, 70, 80, 90, 100

2. Three Hundreds Chart

    They both count by ones. Going down each column skip counts by tens.

Page 13 Practice: Now you try

1. 10, 20, 30, 40, 50, 60, 70, 80, 90, 100
2. 5, 10, 15, 20, 25, 30, 35, 40, 45, 50, 55
3. 550, 555, 560, 565, 570, 575, 580
4. 870, 880, 890, 900, 910, 920, 930
5. 100, 200, 300, 400, 500, 600, 700
6. The numbers always end in 0 or 5.

Ace It Time:

291	292	293	294	295	296	297	298	299	300
301	302	303	304	305	306	307	308	309	310
311	312	313	314	315	316	317	318	319	320
321	322	323	324	325	326	327	328	329	330
331	332	333	334	335	336	337	338	339	340
341	342	343	344	345	346	347	348	349	350
351	352	353	354	355	356				

Start with the highest number, 356, and skip count to the left five numbers. Every fifth number will be called.

356, 351, 346, 341, 336, 331, 326, 321, 316, 311, 306, 301, 296, 291

Show Three-Digit Numbers with Base-Ten Blocks

Page 15 Practice: Now you try

1. 45   2. 369   3. 701   4. 214   5. 503

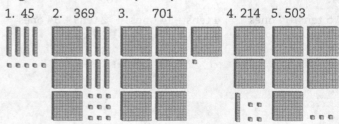

6. 49   7. 84   8. 154   9. 805   10. 667
11. 251   12. 739   13. 893   14. 340   15. 2,450

Ace It Time:

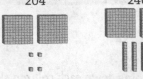

204   240

Both 204 and 240 have two hundreds. 204 doesn't have any tens blocks, but 240 has four. That makes 240 more than 204 because four tens is bigger than no tens.

Expanded and Word Form
Page 17 Practice: Now you try

1.

21 ——— fifty-seven   600 + 70 + 8
57 ——— three hundred forty-two   400 + 8
408 ——— twenty-one   50 + 7
342 ——— six hundred seventy-eight   300 + 40 + 2
678 ——— four hundred eight   20 + 1

2. 729   700 + 20 + 9   seven hundred twenty-nine
3. 503   500 + 0 + 3   five hundred three
4. 956   900 + 50 + 6   nine hundred fifty-six

Ace It Time: In the number 104, or 100 + 0 + 4, there is one hundred and four ones. The zero shows that there are no tens. If you didn't put the 0 in the tens place, the number would be 14. That would show one ten and four ones or 10 + 4.

Show Numbers Different Ways
Page 18 Practice: Now you try

1.	2.	3.
4 tens + 8 ones 40 + 8 48	3 tens + 18 ones 30 + 18 48	2 tens + 28 ones 20 + 28 48

Page 19

4.	5.	6.
5 tens + 4 ones 50 + 4 54	4 tens + 14 ones 40 + 14 54	2 tens + 34 ones 20 + 34 54

Ace It Time: Answers will vary but should illustrate and include each of these combinations:

a. 33 single cupcakes

b. one box of 10 cupcakes + 23 singles

c. two boxes of 10 cupcakes + 13 singles

d. three boxes of 10 cupcakes + 3 singles

Explanations will vary but should indicate an understanding that there are different ways to make a number. The student's explanation should also indicate an understanding that making the number involves adding different combinations of tens and ones to equal the sum of 33.

Compare Three-Digit Numbers
Page 21 Practice: Now you try

1. 673 < 798   673 is less than 798.

2. 453 > 443   453 is greater than 443.

3. 305 > 30 + 5   305 is greater than 35.

4. 599 > 299   599 is greater than 299.

5. You should look at the hundreds place first because that has the largest value.

6. The next place value you look at would be the tens place.

7. 75, 104, 189, 198, 364, 463, 989

Ace It Time: Joey confused his place value. His 101 pennies means he has one hundred, zero tens, and one penny. Joan also has one hundred, but she has one ten. That makes her number of pennies larger than Joey's, because Joey had zero tens.

Mentally Add or Subtract 10 or 100
Page 22 Practice: Now you try

1. 33   2. 15   3. 48   4. 586   5. 753   6. 1,015

Page 23

7. 172   8. 367   9. 003   10. 693   11. 492   12. 1,067

13. 41, 51, 61, 71, 81, 91, 101, 111, 121

14. 1, 101, 201, 301, 401, 501, 601, 701, 801

15. 547, 557, 567, 577, 587, 597, 607, 617

Ace It Time: Lucy walked 665 steps, Mark 365 steps. Since the question asks about how many more *hundreds* of steps Lucy walked, look at the digit in the hundreds place. Lucy walked six sets of 100 steps, and Mark walked three sets of 100 steps. Lucy walked three sets of 100 steps more than Mark, so she walked 300 more steps.

Stop and Think! Unit 2 Review
Page 24 Activity Section 1:

1. 282   2. 346   3. 525   4. 258

Page 25

5. one hundred 6 tens

6. six tens; three hundreds; 360

7. 1 hundred + 6 tens + 2 ones

100 + 60 + 2

Activity Section 2:

1. 474 > 447

Page 26

2. 523 < 535

3. 645, 745

62, 72

55, 60

4. 427, odd

Activity Section 3:

1. 712, even   2. 624, even   3. 705, odd   4. 821, odd

Stop and Think! Unit 2 Understand

Page 27 Activity Section

Ben = 53   Dean = 53   Sarah = 53

Donna = 63   Laura = 53

Donna was incorrect; her way totals 63 since 2 tens = 20 and 43 ones = 43, and 20 + 43 = 63.

Stop and Think! Unit 2 Discover

Page 28 Activity Section

601	602	603	604	605	606	607	608	609	610
611	612	613	614	615	616	617	618	619	620
621	622	623	624	625	626	627	628	629	630
631	632	633	634	635	636	637	638	639	640
641	642	643	644	645	646	647	648	649	650
651	652	653	654	655	656	657	658	659	660
661	662	663	664	665	666	667	668	669	670
671	672	673	674	675	676	677	678	679	680
681	682	683	684	685	686	687	688	689	690
691	692	693	694	695	696	697	698	699	700

1. She bought five groups of 10 rubber bands or 50 rubber bands.

625 + 50 = 675 rubber bands

2. She used six groups of 10 rubber bands or 60 rubber bands.

690 − 60 = 630 rubber bands

Unit 3: Fact Concepts

Addition and Subtraction Mental Fact Strategies

Page 30 Practice: Now you try

1.	6	2.	9	3.	5
	6		10		8
4.	18	5.	13	6.	8
	17		13		9
7.	4	8.	7	9.	10
	7		8		10
10.	6	11.	3	12.	3
	6		2		9
13.	9	14.	1	15.	5
	9		2		4
16.	6	17.	7	18.	6
	4		6		6

Page 31 Ace It Time!

7 (Samantha)   Keegan 7 + 7 = 14 pictures

+ 14 (Keegan)

21 total pictures

Addition and Subtraction Fact Practice

Page 32 Practice: Now you try

1. 11	2. 9	3. 5	4. 4	5. 5	6. 7
7. 4	8. 4	9. 11	10. 10	11. 2	12. 5
13. 15	14. 6	15. 2	16. 9	17. 0	18. 15
19. 7	20. 12	21. 9	22. 17	23. 14	24. 7
25. 4	26. 16	27. 8	28. 9	29. 13	30. 10
31. 6	32. 6	33. 17	34. 13	35. 8	36. 16
37. 4	38. 2	39. 2	40. 9	41. 18	42. 8

Page 33 Ace It Time!

9 (pages read on Monday)   20 (pages in book)

+ 7 (pages read on Tuesday)   − 16 (pages read)

16 total pages read   4 more pages to read

Addition and Subtraction Fact Families

Page 34 Practice: Now you try

1. 8 + 2 = 10   2 + 8 = 10   10 − 8 = 2   10 − 2 = 8

2. 6 + 3 = 9   3 + 6 = 9   9 − 6 = 3   9 − 3 = 6

3. 7 + 8 = 15   8 + 7 = 15   15 − 7 = 8   15 − 8 = 7

4. 5 + 6 = 11   6 + 5 = 11   11 − 5 = 6   11 − 6 = 5

5. 4 + 2 = 6   2 + 4 = 6   6 − 4 = 2   6 − 2 = 4

6. 5 + 12 = 17   12 + 5 = 17   17 − 5 = 12   17 − 12 = 5

Answer Key

7. 9   8. 11   9. 16
   5   7   9

10. 14   11. 8   12. 18
   8   3   9

13. 12   14. 10   15. 13
   3   4   8

Ace It Time:

     7 (bag 1)   15 (total apples)
   + 8 (bag 2)   – 6 (apples eaten)
    15 apples   9 apples left

Addition Using Arrays

Page 36 Practice: Now you try

1. Check student's drawing. Should represent three rows of two.

Solve:

$2 + 2 + 2 = 6$

$3 + 3 = 6$

= 6 pictures

2. Check student's drawing. Should represent five rows of five.

Solve:

$5 + 5 + 5 + 5 + 5 = 25$

= 25 books

Page 37

3. Check student's drawing. Should represent four rows of five.

Solve:

$5 + 5 + 5 + 5 = 20$

$4 + 4 + 4 + 4 + 4 = 20$

= 20 stamps

4. Check student's drawing. Should represent three rows of six.

Solve:

$6 + 6 + 6 = 18$

$3 + 3 + 3 + 3 + 3 + 3 = 18$

= 18 eggs

5. Check student's drawing. Should represent two rows of eight.

Solve:

$8 + 8 = 16$

$2 + 2 + 2 + 2 + 2 + 2 + 2 + 2 = 16$

= 16 candles

Ace It Time:

Check student's drawing. Should represent four rows of six. Then represent subtracting two crayons.

Solve:

$6 + 6 + 6 + 6 = 24$

$4 + 4 + 4 + 4 + 4 + 4 = 24$

   24 crayons

– 2 broken crayons

   22 crayons not broken

Unit 4: Addition and Subtraction Concepts

Addition Using Place Value

Page 38 Practice: Now you try

1. $20 + 1$

$40 + 6$

$60 + 7 = 67$

Page 39

2. $30 + 6$

$20 + 2$

$50 + 8 = 58$

3. $10 + 5$

$50 + 3$

$60 + 8 = 68$

4. $20 + 4$

$30 + 4$

$50 + 8 = 58$

5. $60 + 3$

$10 + 2$

$70 + 5 = 75$

Ace It Time:

Sondra   22 crayons $20 + 2$

Meagan   + 16 crayons $10 + 6$

                        $30 + 8 = 38$ crayons in all

Addition Using the Arrow Method

Page 40 Practice: Now you try

Arrows will vary in each problem. Check arrows in each problem for accuracy.

1. 85
2. 69
3. 96
4. 75
5. 59
6. 107

Page 41

7. 103

8. 58

9. 87

10. 67

Ace It Time:

Michael	34	Martin	33
Martin	+ 33	Micah	+ 35
	67 (not correct)		68 (not correct)

Michael	34
Micah	+ 35
	69 (correct)

The two players whose ticket sales equaled 69 tickets were Michael and Micah.

Make a Ten and Add On

Page 42 Practice: Now you try

1. 13: 10 + 3 = 13

2. 13: 10 + 3 = 13

3. 11: 10 + 1 = 11

4. 11: 10 + 1 = 11

5. 17: 10 + 7 = 17

6. 14: 10 + 4 = 14

7. 12: 10 + 2 = 12

8. 11: 10 + 1 = 11

9. 11: 10 + 1 = 11

10. 14: 10 + 4 = 14

Page 43 Ace It Time:

6 + 7 = 13   6 + 9 = 15 The sum between these numbers is 14; the only doubles fact that equals 14 is 7 + 7.

Regroup Using Place Value

Page 45 Practice: Now you try

1. 40 + 8	2. 30 + 9	3. 60 + 7
20 + 6	50 + 4	10 + 5
60 + 14	80 + 13	70 + 12
60 + 10 + 4 = 74	80 + 10 + 3 = 93	70 + 10 + 2 = 82
4. 40 + 9	5. 30 + 7	
20 + 3	50 + 5	
60 + 12	80 + 12	
60 + 10 + 2 = 72	80 + 10 + 2 = 92	

Page 46 Ace It Time:

Juan's crystals   48

Friends' crystals + 26

40 + 8

20 + 6

60 + 14

60 + 10 + 4 = 74 crystals

Add Numbers in a Series

Page 48 Practice: Now you try

1. 78

2. 89

3. 88

4. 129

5. 91

6. 64

7. 138

8. 85

9. 88

10. 100

11. 118

12. 73

Page 49 Ace It Time:

The number that is the sum of 7 and 7: 14

The number between 30 and 40:   35

14	
+ 35	
49	49 + _____ = 97

The only numbers left are 16, 17, 25, and 48.

49	49	49	49
+ 16	+ 17	+ 25	+ 48
65	66	74	97

The three numbers are 14, 35, and 48.

Add Three-Digit Numbers

Page 51 Practice: Now you try

1. 934

2. 496

3. 811

4. 701

5. 620

6. 972

7. 706

8. 455

9. 462

10. 875

11. 895

12. 601

Ace It Time:

The two students were Juan and Juanita

Juan	327	Juan	327	Julio	237
Juanita	+ 273	Julio	+ 237	Juanita	+ 273
	600		564		510

Subtraction Using Place Value

Pages 52–53 Practice: Now you try

1. 70 + 9	2. 50 + 6	3. 60 + 8
− 40 + 6	− 10 + 4	− 30 + 7
30 + 3	40 + 2	30 + 1
30 + 3 = 33	40 + 2 = 42	30 + 1 = 31
4. 40 + 5	5. 90 + 3	
− 20 + 3	− 70 + 1	
20 + 2	20 + 2	
20 + 2 = 22	20 + 2 = 22	

Answer Key

Ace It Time:

Toys in the box to begin: 57

Toys in there now: − 34

                                23 toys taken out of the box

50 + 7

− 30 + 4

20 + 3 = 23

Add Up to Subtract

Page 54 Practice: Now you try

Number lines will vary depending on the increments chosen to add up by. Check number lines for accuracy.

1. 53 − 45 = 8 (Possible answer: Start at 45 + 5 = 50: then 50 + 3 = 53: so 5 + 3 = 8)

2. 75 − 68 = 7 (Possible answer: Start at 68 + 2 = 70; then 70 + 5 = 75: so 2 + 5 = 7)

3. 34 − 29 = 5 (Possible answer: Start at 29 + 1 = 30: then 30 + 4 = 34: so 1 + 4 = 5)

4. 97 − 88 = 9 (Possible answer: Start at 88 + 2 = 90: then 90 + 7 = 97: so 7 + 2 = 9)

Page 55

5. 79 − 63 = 16 (Possible answer: Start at 63 + 7 = 70: then 70 + 9 = 79: so 7 + 9 = 16)

6. 58 − 41 = 17 (Possible answer: Start at 41 + 9 = 50: then 50 + 8 = 58: so 9 + 8 = 17)

Ace It Time:

22 − 7 = 15: Christine solved 15 math problems.

Sample answer on the number line could be: Start at 22 then count backward by 1s seven times until you get to the number 15.

Subtraction Using the Arrow Method

Page 56 Practice: Now you try

Arrow methods will vary depending on the increments the student chose to subtract by.

1. 15
2. 14
3. 33
4. 16
5. 10
6. 31

Page 57

7. 34
8. 35
9. 42
10. 21

Ace It Time:

96 math papers	Another way:
− 35 Monday	Monday 35 papers graded
61 papers left on Monday	Tuesday + 35 papers graded
− 35 Tuesday	70
26 papers left to grade on Wednesday	Math papers to be graded 96
	Papers graded Mon/Tues − 70
	26 left

Subtract Two-Digit Numbers with Regrouping

Page 58 Practice: Now you try

1.
5 tens	12 ones
− 3 tens	7 ones
2 ten	5 ones

	6	2
−	3	7
	2	5

2.
6 tens	11 ones
− 6 tens	4 ones
0	7 ones

	7	1
−	6	4
		7

3.
4 tens	16 ones
− 2 tens	7 ones
2 tens	9 ones

	5	6
−	2	7
	2	9

4.
5 tens	13 ones
− 4 tens	5 ones
1 ten	8 ones

	6	3
−	4	5
	1	8

Page 59 Practice: Now you try

5. 16   6. 32   7. 38   8. 8   9. 46

10. 8   11. 35   12. 15   13. 39   14. 18

Ace It Time:

24 French paintings	75 total paintings
+ 51 American paintings	− 37 photographs
75 total paintings	38 more paintings

Subtract Three-Digit Numbers with Regrouping

Page 60 Practice: Now you try

1. 277   7. 204
2. 75   8. 281
3. 267   9. 359
4. 483   10. 235
5. 522   11. 256
6. 97   12. 99

Page 61

13. 848   14. 856   15. 823
    − 361   − 728   − 148
     487   128   675

Ace It Time:

500   I am the number 253.

− 247

253

Stop and Think! Units 3–4 Review
Page 62 Activity Section 1:

1. 6 + 3 = <u>9</u>	2. 7 – 4 = <u>3</u>	3. 2 + 8 = <u>10</u>
3 + 6 = 9	7 – 3 = 4	8 + 2 = 10
9 – 3 = 6	3 + 4 = 7	10 – 2 = 8
9 – 6 = 3	4 + 3 = 7	10 – 8 = 2
4. 12 – 9 = <u>3</u>	5. 7 + 8 = <u>15</u>	6. 14 – 6 = <u>8</u>
12 – 3 = 9	8 + 7 = 15	14 – 8 = 6
9 + 3 = 12	15 – 8 = 7	8 + 6 = 14
3 + 9 = 12	15 – 7 = 8	6 + 8 = 14

7. 12	8. 3	9. 5	10. 12	11. 6	12. 6	13. 0
14. 5	15. 5	16. 2	17. 6	18. 2	19. 10	20. 9
21. 10	22. 3	23. 10	24. 3	25. 8	26. 14	27. 10

Page 63 Activity Section 2:

1. 75	2. 31	3. 923	4. 624	5. 117
6. 85	7. 96	8. 956	9. 679	10. 732

Activity Section 3:

1. 2	2. 6	3. 5	4. 1	5. 4
6. 0	7. 4	8. 10	9. 1	10. 2

11. 18	12. 30	13. 33	14. 14	15. 14
16. 213	17. 356	18. 513	19. 119	20. 68

Page 64 Activity Section 4:

1. 938 Kaitlyn's pennies	2. 256 miles on day one
– 900 pennies taken to the bank	+ 391 miles on day two
38 pennies left	647 miles in all

Stop and Think! Units 3–4 Understand
Page 65

Explanations will vary but should indicate the following mistakes.

Chloe added 62 + 20 wrong. 62 + 20 = 82 not 72.	Charley was correct.
Kyle took 8 from the 28 and added it correctly to 62 to make the ten of 70. Instead of taking the 8 away from the 28, he added 2 to 28 to make another ten.	Ginger regrouped the sum of 2 + 8 = 10 incorrectly. She put the 1 down in the ones place and then carried the 0 over to the tens place.

Stop and Think! Units 3–4 Discover
Page 66

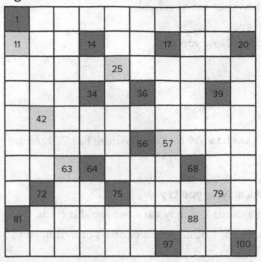

Explanations and calculations will vary but should include adding or subtracting numbers from the given numbers.

For example: 11 – 10 = 1; 11 + 3 = 14; 14 + 3 = 17; 17 + 3 = 20; 14 + 20 = 34; 34 + 2 = 36; 36 + 3 = 39; 56 + 1 = 57; 63 + 1 = 64; 64 + 4 = 68; 42 + 30 = 72; 72 + 3 = 75; 11 + 70 = 81; 88 + 9 = 97; 97 + 3 = 100.

Unit 5: Money and Time Concepts
What Time Is It?
Page 68 Practice: Now you try

1. 8:10
2. 6:40
3. 3:25
4. 9:50
5. Long hand on the 2 and short hand between the 7 and 8
6. Long hand on the 7 and short hand between the 10 and 11
7. Long hand on the 9 and short hand between the 4 and 5

Ace It Time: The time on the clock shows 3:40.

A.M. or P.M.?
Page 69 Practice: Now you try

1. 4:15 P.M.
2. 11:40 A.M.
3. 5:55 P.M.
4. 7:10 A.M.

Page 70

5. Long hand on the 9 and short hand between the 1 and 2, P.M.
6. Long hand on the 6 and short hand between the 8 and 9, P.M.

Answer Key

Ace It Time: Matthew is correct. 2:00 A.M. is after midnight; the boys should be asleep. 2:00 P.M. is in the afternoon, when they are at school.

Dollars and Cents

Page 71 Practice: Now you try

1. 64¢

Page 72

2. $2.97

3. 86¢

Ace It Time: Kara has 61¢ cents and Andre has 72¢. Andre has 11¢ more.

Make Money Different Ways

Page 74 Practice: Now you try

Answers to questions 1–7 may vary but could include:

1. 63¢ = 2 quarters, 1 dime, and 3 pennies or 6 dimes and 3 pennies
2. 78¢ = 3 quarters and 3 pennies or 2 quarters, 2 dimes, 1 nickel, and 3 pennies
3. 29¢ = 1 quarter and 4 pennies or 2 dimes, 1 nickel, and 4 pennies
4. 55¢ = 2 quarters and 5 pennies or 5 dimes and 1 nickel
5. 42¢ = 1 quarter, 3 nickels, and 2 pennies or 4 dimes and 2 pennies
6. 66¢ = 2 quarters, 1 dime, and 6 pennies or 4 dimes, 4 nickels, and 6 pennies
7. 33¢ = 1 quarter, 1 nickel, and 3 pennies or 2 dimes, 2 nickels, and 3 pennies

Page 75

Ace It Time: Megan has 93¢. She can also make that with 9 dimes and 3 pennies.

Unit 6: Measurement and Data Concepts

Measure the Length

Page 77 Practice: Now you try

1. 3 inches
2. 5 inches
3. 6 inches
4. 2 inches
5. 10 centimeters
6. 18 centimeters

Page 78

7. 6 centimeters
8. 13 centimeters

Ace It Time: Martez should use a yardstick because a driveway is a large object. A yardstick is used to measure large objects.

Guess the Length

Page 79 Practice: Now you try

1. 2 inches
2. 3 inches

Page 80

3. 6 centimeters
4. 8 centimeters
5. 16 centimeters
6. 12 inches

Ace It Time: I estimate that the cell phone and mail envelope measure closer to 15 inches in length. My notebook measures a little over 8 inches by 11 inches. A magazine is about the same size as or bigger than my notebook. So together, those two things will be more than 15 inches in length.

Measure the Same Length with Different Units

Page 81 Practice: Now you try

1. 4 inches, 10 centimeters; So it takes more centimeters than inches to measure the nail.
2. 5 inches, 12 centimeters; So it takes more centimeters than inches to measure the water bottle.

Page 82

3. 8 inches, 20 centimeters; So it takes more centimeters than inches to measure the book.
4. centimeters
5. yards

Ace It Time: Zen's measurements are wrong because a centimeter is a smaller unit than an inch, so there should be more centimeters when measuring an object with both centimeters and inches.

Compare Lengths

Page 83 Practice: Now you try

1. The top piece of licorice is 3 inches long. The bottom piece of licorice is 5 inches long.

   5 inches – 3 inches = 2 inches. The top piece of licorice is 2 inches shorter than the bottom piece.

Page 84

2. The fork is 14 centimeters long. The knife is 17 centimeters long.

   17 centimeters – 14 centimeters = 3 centimeters. The fork is 3 centimeters shorter than the knife.

Ace It Time: Katie's mop needs to be 14 inches longer to be the same length as the broom because 62 inches – 48 inches = 14 inches.

Solve Problems by Adding and Subtracting Lengths

Page 86 Practice: Now you try

1. 22 inches
2. 25 meters
3. 16 yards

Ace It Time: The lamp in Mrs. Miller's dining room is 67 inches because 55 inches + 12 inches = 67 inches.

Picture Graphs

Page 87 Cookies for the Party

Cookies for the Party	
Type of Cookie	**Number of Cookies**
Chocolate Chip	4
Peanut Butter	3
Oatmeal Raisin	5
Sugar	3

Page 88 Practice: Now you try

1.

Books Derrick Read	
Month	**Number of Books**
May	4
June	6
July	8
August	5

2.

Books Derrick Read	
Month	**Number of Books**
May	
June	
July	
August	

Key: Each 📖 = 1 book

3. May

4. July

5. 4 more books

6. 13 books

Page 89

Ace It Time: There are three more cookies. Oatmeal raisin and chocolate chip cookies combined = 9. Peanut butter and sugar cookies combined = 6. 9 − 6 = 3

Bar Graphs

Page 90 Practice: Now you try

Step 1

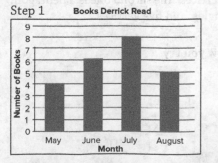

Books Derrick Read

Page 91

Step 2

1. July

2. May

3. 3 more books

4. 12 books

Ace It Time: Derrick read three more books in the second part of the summer, because the total number of books he read in the second part of summer is 13 (8 + 5 = 13). The first part of summer he read 10 books (4 + 6 = 10). To find out how many more, we subtract 13 − 10 = 3.

Line Plots

Pages 93 and 94 Practice: Now you try

Step 1

Length of Time Students Walked in Mrs. Ortega's Class	
Time (in Minutes)	**Number of Students**
14	3
15	4
16	2
17	2

Steps 2 and 3

Title: (Possible Title) Time it Took Students to Walk One Mile

```
                    X
         X    X
         X    X    X    X
         X    X    X    X
    ◄────┼────┼────┼────┼────►
         14   15   16   17
              Time (in Minutes)
```

1. 11

2. two more students

3. eight students

4. four students

Ace It Time: Since seven students walked for 15 minutes or less and four students walked for 16 minutes or more, then the number of students that walked for 15 minutes or less is greater.

Stop and Think! Units 5–6 Review

Page 95

Activity Section 1:

1. 3:45 P.M.

2. Long hand on the 3 and short hand between the 8 and 9, P.M.

Activity Section 2:

1. 51¢

2. $1.44

Page 96

3. Answers to question 3 may vary but could include: $1.34 = 4 quarters, 3 dimes, and 4 pennies or 13 dimes and 4 pennies

Activity Section 3:

1. 5 inches

2. 9 centimeters

3. 3 inches

4. 6 centimeters

5. centimeters

6. yards

Page 97

Activity Section 4:

1. 21 + 27 = 48 inches

Activity Section 5:

1. 8

2. 4

3. 12

4. 3

5. Lilly and Melissa

6. 9

7. 6

8. 3

9. 5

Stop and Think! Units 5–6 Understand

Page 98

1. 14 choices were not chocolate.

Step 1

Favorite Ice Cream Flavor	
Ice Cream Flavor	Number of Students
Chocolate	5
Strawberry	7
Vanilla	3
Cookies and Cream	4

Steps 2 and 3

Step 1

Number of Jumping Jacks Done in Mrs. Hoffman's Class	
Number of Jumping Jacks	Number of Students
24	3
25	5
26	2
27	2

Steps 2 and 3

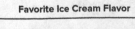

Number of Jumping Jacks Done in Mrs. Hoffman's Class
X
X
X   X
X   X   X   X
X   X   X   X
24   25   26   27

2. 3

3. 3

4. 4

Stop and Think! Units 5–6 Discover

Page 99

1. Che has 45 cents left because 73 cents – 28 cents = 45 cents. The least number of coins he could have to make 45 cents is three coins (1 quarter and 2 dimes).

2. The most amount of coins Che could have to make 45 cents is 45 coins, or 45 pennies.

Unit 7: Geometry Concepts

Polygons and Shapes

Page 101 Practice: Now you try

1. quadrilateral
   sides: 4
   angles: 4

2. triangle
   sides: 3
   angles: 3

3. hexagon
   sides: 6
   angles: 6

4. pentagon
   sides: 5
   angles: 5

Ace It Time:

Same: Both are polygons. Both have angles and vertices.

Different: The first one is a quadrilateral. It has four sides, four angles, and four vertices. The second one is a pentagon. It has five sides, five angles, and five vertices.

The Cube

Page 102 Practice: Now you try

1.

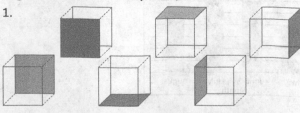

Page 103

2.

Ace It Time: Jaxon is correct; only one of the shapes is a cube even though they both have six faces, eight vertices, and 12 edges. A cube must have squares for all six faces. The second shape is a cube. The first is not because it does not have six squares as faces. Four of the faces on that shape are rectangles, so it cannot be a cube.

Let's Draw Polygons
Page 104 Practice: Now you try

1. Triangle — Sides: 3, Angles: 3
2. Quadrilateral — Sides: 4, Angles: 4
3. Pentagon — Sides: 5, Angles: 5
4. Hexagon — Sides: 6, Angles: 6

5. 6. 7.

Page 105

8.

Ace It Time: Sample Answers

Pentagon—main shape
Quadrilateral—1 and 2
Triangles—3 and 4
Hexagon—5

Triangles—Main shape and front part of flag
Quadrilaterals—two stripes and back part of flag
Hexagon—1
Pentagon—2

The Great Rectangle Divide
Page 107 Practice: Now you try

1.
_3_ squares   _1_ row   _3_ columns

2.
_10_ squares   _2_ rows   _5_ columns

Ace It Time: The student should draw a rectangle similar to this sample with two rows and three columns.

Unit 8: Fraction Concepts
What Is a Fraction?
Page 109 Practice: Now you try

1. and 2.
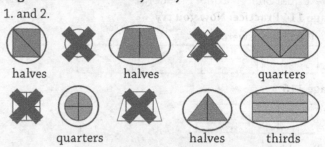

halves   halves   quarters
quarters   halves   thirds

Ace It Time: Zia cut the sandwich into three equal pieces because cutting something into thirds means three equal pieces.

Name the Fraction
Page 110 Practice: Now you try

1. 2.
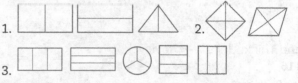
3.

Page 111

Name the fractions. Tell how many equal parts.

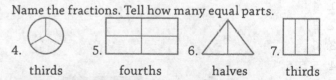

4. thirds — 3 equal parts
5. fourths — 4 equal parts
6. halves — 2 equal parts
7. thirds — 3 equal parts

Ace It Time: Neal forgot that he needed a piece. He cut the cake into thirds for three of his friends. He should have cut the cake into fourths, because there were three friends plus Neal, which makes four.

Fraction Action with Rectangles and Circles
Page 112 Practice: Now you try

1. Shade one-third. 2. Shade one-fourth. 3. Shade one-half.

Page 113

4. _3_ thirds to make one whole. Draw a picture to show this.
5. _2_ halves to make one whole. Draw a picture to show this.
6. _4_ fourths to make one whole. Draw a picture to show this.

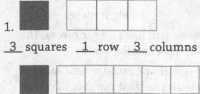
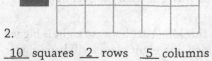
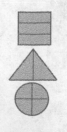

Answer Key

Ace It Time: Clay ate one piece and each of his friends ate one piece.

They ate six pieces in all. Since there were only six pieces, they ate the whole cake.

Are Equal Shares Equal Fractions?

Page 114 Practice: Now you try

1. Possible answers

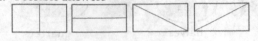

Page 115

2.

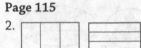

3. Possible answers

Ace It Time:

Stop and Think! Units 7–8 Review

Page 116

Activity Section 1:

1. 4
2. 3
3. 6
4. 5, 5, 5

5. triangle 6. pentagon 7. quadrilateral (rectangle)

8. 12 edges
9. 6 faces
10. 8 vertices

Page 117 Activity Section 2:

1. 3
2. 2
3. 6
4. 8 squares   5. 12 squares

Page 118 Activity Section 3:

1.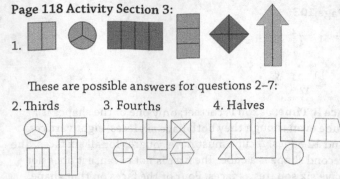

These are possible answers for questions 2–7:

2. Thirds   3. Fourths   4. Halves

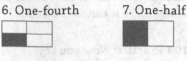

5. One-third   6. One-fourth   7. One-half

8.

Stop and Think! Units 7–8 Understand

Page 119

Activity Section

Complete the table

Number of Items	Fractional Part	Picture	How Much Is the Fraction?
12 eggs	one-third	x x x x / x x x x / x x x x	one-third of 12 = 4
10 marbles	one-half	x x x x x / x x x x x	one-half of 10 = 5
15 raisins	one-third	x x x x x / x x x x x / x x x x x	one-third of 15 = 5
12 apples	one-fourth	x x x / x x x / x x x / x x x	one-fourth of 12 = 3

Stop and Think! Units 7–8 Discover

Page 120

Activity Section

six pieces   eight pieces

The pizza with only six pieces will have bigger pieces. The pizza with eight pieces will have smaller pieces.

When you take the same-size shape and divide it into smaller pieces, the shape with the least number of pieces will have bigger parts.

Martin is wrong. He is thinking that because eight is bigger than six, the pizza divided into eight pieces will have bigger pieces.